NATURAL DISASTER RESEARCH, PREDICTION AND MITIGATION

FEMA'S DISASTER RELIEF FUND

ELEMENTS AND CONSIDERATIONS

NATURAL DISASTER RESEARCH, PREDICTION AND MITIGATION

Additional books in this series can be found on Nova's website under the Series tab.

Additional e-books in this series can be found on Nova's website under the e-book tab.

NATURAL DISASTER RESEARCH, PREDICTION AND MITIGATION

FEMA'S DISASTER RELIEF FUND

ELEMENTS AND CONSIDERATIONS

ANGELICA CONNER
EDITOR

NOVINKA
New York

Library of Congress Cataloging-in-Publication Data

ISBN: 978-1-63321-630-3

Published by Nova Science Publishers, Inc. † New York

CONTENTS

Preface

Chapter 1 – The Robert T. Stafford Emergency Relief and Disaster Assistance Act authorizes the President to issue declarations for incidents ranging from destructive, large-scale disasters to more routine, less damaging events. Declarations trigger federal assistance in the forms of various response and recovery programs under the Stafford Act to state, local, and tribal governments. The Federal Emergency Management Agency's (FEMA) Disaster Relief Fund (DRF) is the primary funding source for disaster response and recovery.

This report describes the declaration process and the types of declarations that can be issued under the Stafford Act: (1) emergency and major disaster declarations, and (2) Fire Management Assistance Grants. The report also examines how the DRF is financed. This discussion is followed by an analysis concerning the issues related to the DRF including the debate over supplemental appropriations, how the DRF is budgeted, and the influence the Budget Control Act has had on the DRF.

Chapter 2 – When a state is overwhelmed by an emergency or disaster, the governor may request assistance from the federal government. Federal assistance is contingent on whether the President issues an emergency or major disaster declaration. Once the declaration has been issued the Federal Emergency Management Agency (FEMA) provides disaster relief through the use of the Disaster Relief Fund (DRF), which is the source of funding for the Robert T. Stafford Emergency Relief and Disaster Assistance Act response and recovery programs. Congress appropriates money to the DRF to ensure that funding for disaster relief is available to help individuals and communities stricken by emergencies and major disasters (in addition, Congress appropriates disaster funds to other accounts administered by other federal agencies pursuant to federal statutes that authorize specific types of disaster relief).

Historically, the DRF is generally funded at a level that is sufficient for what are known as "normal" disasters. These are incidents for which DRF outlays are less than $500 million. When a large disaster occurs, funding for the DRF may be augmented through emergency supplemental appropriations. A supplemental appropriation generally provides additional budget authority during the current fiscal year to (1) finance activities not provided for in the regular appropriation; or (2) provide funds when the regular appropriation is deemed insufficient.

This methodology used to budget the DRF appears to have been altered since the passage of the Budget Control Act (P.L. 112-25). The Budget Control Act includes a series of provisions that directed the Office of Management and Budget (OMB) to annually calculate an "allowable adjustment" for disaster relief to the BCA's discretionary spending caps. That adjustment, if used, would make additional budget authority available for the federal costs incurred by major disasters declared under the Stafford Act beyond what is allowed in the regular discretionary budget allocation. The OMB calculation may have provided a mechanism that encourages a larger regular appropriation to the DRF. It is possible that larger DRF appropriations may reduce the need for supplemental appropriations.

Budgeting for disaster relief has been the subject of a great deal of debate. Some argue that more money should be appropriated in FEMA's DRF account in annual appropriations, while others maintain that augmenting the DRF through supplemental appropriations is preferable because it allows Congress to react directly to a particular situation. Others may argue that emergency supplemental appropriations are preferable for fiscal management reasons because an appropriation is not requested unless there is a real need for supplemental funding. Another argument is to revamp the budgetary process to fund disaster relief.

This report describes the various components of the DRF, including (1) what authorities have shaped it over the years; (2) how FEMA determines the amount of the appropriation requested to Congress (pertaining to the DRF); and (3) how emergency supplemental appropriations are requested. Information is also provided on funds appropriated in supplemental appropriations legislation to agencies other than the Department of Homeland Security (DHS). Aspects of debate concerning how disaster relief is budgeted are also highlighted and examined, and alternative budgetary options are summarized.

Chapter 3 – The Robert T. Stafford Disaster Relief and Emergency Assistance Act (referred to as the Stafford Act - 42 U.S.C. 5721 et seq.) authorizes the President to issue "major disaster" or "emergency" declarations before or after catastrophes occur. Emergency declarations trigger aid that protects property, public health, and safety and lessens or averts the threat of an incident becoming a catastrophic event. A major disaster declaration, issued after catastrophes occur, constitutes broader authority for federal agencies to provide supplemental assistance to help state and local governments, families and individuals, and certain nonprofit organizations recover from the incident.

The end result of a presidential disaster declaration is well known, if not entirely understood. Various forms of assistance are provided, including aid to families and individuals for uninsured needs and assistance to state and local governments and certain non-profits in rebuilding or replacing damaged infrastructure. The amount of assistance provided through presidential disaster declarations has exceeded $140 billion. Often, in recent years, Congress has enacted supplemental appropriations legislation to cover unanticipated costs. While the amounts spent by the federal government on different programs may be reported, and the progress of the recovery can be observed, much less is known about the process that initiates all of this activity. Yet, it is a process that has resulted in an average of more than one disaster declaration a week over the last decade.

The disaster declaration procedure is foremost a process that preserves the discretion of the governor to request assistance and the President to decide to grant, or not to grant, supplemental help. The process employs some measurable criteria in two broad areas: Individual Assistance that aids families and individuals and Public Assistance that is mainly for repairs to infrastructure. The criteria, however, also considers many other factors, in each category of assistance, that help decision makers assess the impact of an event on communities and states. Under current law, the decision to issue a declaration rests solely with the President. Congress has no formal role, but has taken actions to adjust the terms of the process. For example, P.L. 109-295 established an advocate to help small states with the declaration process. More recently, Congress introduced legislation, H.R. 3377, that would direct FEMA to update some of its criteria for considering Individual Assistance declarations. Congress continues to examine the process and has received some recommendations for improvements. Given the importance of the decision, and the size of the overall spending involved, hearings have been held to review the declaration process so as to ensure fairness and equity in the process and its results.

In: FEMA's Disaster Relief Fund
Editor: Angelica Conner
ISBN: 978-1-63321-630-3

Chapter 1

FEMA's Disaster Relief Fund: Overview and Selected Issues*

Bruce R. Lindsay

Summary

The Robert T. Stafford Emergency Relief and Disaster Assistance Act (P.L. 93-288, as amended) authorizes the President to issue declarations for incidents ranging from destructive, large-scale disasters to more routine, less damaging events. Declarations trigger federal assistance in the forms of various response and recovery programs under the Stafford Act to state, local, and tribal governments. The Federal Emergency Management Agency's (FEMA) Disaster Relief Fund (DRF) is the primary funding source for disaster response and recovery.

Funds from the DRF are used to pay for ongoing recovery projects from disasters occurring in previous fiscal years, meet current emergency requirements, and as a reserve to pay for upcoming incidents. The DRF is funded annually and is a "no-year" account, meaning that unused funds from the previous fiscal year (if available) are carried over to the next fiscal year. In general, when the balance of the DRF becomes low, Congress provides

* This is an edited, reformatted and augmented version of a Congressional Research Service publication R43537, prepared for Members and Committees of Congress dated May 7, 2014.

additional funding through both annual and supplemental appropriations to replenish the account.

The federal government provides a significant amount of money to state and local governments each year for emergency and major disasters. For example, Congress provided roughly $120 billion for Hurricane Katrina and $60 billion for Hurricane Sandy recovery. Even in years with relatively few major disasters, it is not uncommon for the federal government to annually appropriate between $2 billion and $6 billion to help pay for recovery projects. Studies and analyses of disasters indicate that there has been an uptick in the number of major disasters declared each year. In addition, scholars of disaster policy and other experts such as climatologists expect disasters to increase in both frequency and in costs in the near future.

Federal disaster assistance expenditures are influenced by both external and internal factors. External factors that increase federal spending on disaster costs include increases in the frequency and magnitude of weather related events, and increases in population size and development— especially in coastal and other flood prone areas. Internal factors also influence how much assistance is provided and include disaster assistance policies that have evolved over time that have expanded the federal role in emergency and major disaster declarations such as altering declaration criteria and adjusting the federal cost-share for response and recovery.

Congressional interest in disaster assistance has always been high given the amount of money provided to states and localities, but also because of increasing disagreements over the appropriate role of the federal government in providing assistance. Other congressional concerns include the use of supplemental appropriations to pay for disaster relief, offsetting expenditures for disaster assistance, and whether some of the federal burden for disaster assistance should be shifted to states and localities.

This report describes the declaration process and the types of declarations that can be issued under the Stafford Act: (1) emergency and major disaster declarations, and (2) Fire Management Assistance Grants. The report also examines how the DRF is financed. This discussion is followed by an analysis concerning the issues related to the DRF including the debate over supplemental appropriations, how the DRF is budgeted, and the influence the Budget Control Act has had on the DRF.

Some argue that the current method of funding and providing federal assistance for disaster response and recovery is functioning correctly and should not be changed. Others argue that the federal government should increase the amount of funding provided to states and localities for emergency and major disaster declarations. Still others argue that policy options that reduce federal costs for emergency and major disaster declarations or reduce the number of supplemental appropriations needed (or both) should be pursued. Policy proposals that could help achieve these ends include:

- appropriating more funds for the DRF to reduce the need for supplemental funding,
- restructuring the budget procedures for disaster assistance,
- creating alternative funding methods such as a rainy-day fund or a contingency fund,
- reducing federal costs by eliminating unrelated spending in disaster funding bills,
- altering policies that would limit the number of declarations issued each year, and
- converting some or all disaster assistance to disaster loans.

This report concludes with policy questions that may help frame future discussions concerning federal emergency and disaster relief.

OVERVIEW OF THE DISASTER RELIEF FUND

The Disaster Relief Fund (DRF), sometimes referred to as the President's Disaster Relief Fund, is managed by the Federal Emergency Management Agency (FEMA).[1] The DRF is the main account used to fund a wide variety of disaster assistance programs that provide grants and other forms of support to assist state local, and tribal governments, as well as certain nonprofit entities during disaster recovery. The DRF is also used to fund Mission Assignments. Mission Assignments are used by FEMA to task and reimburse other federal entities that provide direct assistance during emergency and major disaster declarations.[2] The DRF functions as a reserve for potential, future incidents, as well as an account to pay for ongoing projects to recover from past disasters.

The majority of assistance provided by the federal government in response to emergency and major disaster declarations is funded through the DRF. The DRF, however, is not used to fund the federal response to every type of incident. In general, funds from the DRF are released after the President has issued an emergency or major disaster declaration pursuant to the Robert T. Stafford Relief and EmergencyAssistance Act (P.L. 93-288, as amended, hereinafter the Stafford Act).[3] Incidents that are not declared under the

Stafford Act are either handled by states and localities without federal assistance, or the assistance is provided by another federal entity under its own authority using its own funding mechanism. For example, the Small BusinessAct[4] authorizes the Small Business Administration (SBA) to issue disaster loans to households and businesses in response to certain types of incidents.[5] The U.S. Department of Agriculture (USDA) also offers several permanently authorized programs to help farmers recover financially from a natural disaster, including federal crop insurance, the Noninsured Crop Disaster Assistance Program (NAP), and emergency disaster loans.[6]

While not specifically authorized in statute, the DRF has been used to fund recovery projects for over four decades. Rather than an authorized account, the DRF is the product of legislation and federal policies that can be traced to the post-World War II era. Prior to that time, disaster response activities were funded primarily through local efforts and voluntary groups. In cases where the federal government did offer assistance, the needs of disaster victims and affected communities were funded on an as-needed basis through appropriations that were then allocated, pursuant to the legislation, by executive branch administrators and, ultimately, the President.

Presidential Declarations

There are two principal forms of presidential action that authorize federal assistance to states and localities that are paid out of the DRF: emergency declarations and major disaster declarations.

- Emergency declarations are issued to protect property and public health and safety, and to lessen or avert the threat of a major disaster or catastrophe.[7] Emergency declarations are often made when a threat is recognized (for example, before a hurricane makes landfall) and are intended to supplement and aid the coordination of state and local efforts prior to the event, such as evacuations and protection of public assets.
- Major disaster declarations, on the other hand, are generally made as a result of a significant incident and constitute a broader authority that helps states and local communities, as well as families and individuals, recover from the damage caused by the event.[8]

The Stafford Act Process for Declaring Emergencies and Major Disasters

As shown in **Figure 1**, the declaration process used by the federal government is based on the concept of scalability. For example, suppose communities and local governments are the first to respond after an incident has occurred. Local governments must then request assistance from the state if responding to (or recovering from) the incident is beyond their capacity. Similarly, when a state is overwhelmed by an incident, the state governor may elect to request assistance from the federal government. Federal assistance is contingent on the gubernatorial request because the Stafford Act stipulates that the governor of an affected state must formally ask the President to issue an emergency or major disaster declaration.

A President's declaration triggers the allocation of funds from the DRF, and the funding may be distributed from any one, or any combination, of three categories of disaster aid:

1. **Individual Assistance.** Individual Assistance (IA) includes disaster housing for displaced individuals, grants for needs not covered by insurance, crisis counseling, and disaster-related unemployment assistance.
2. **Public Assistance.** Public assistance (PA), which is FEMA's largest funded program, helps communities absorb the costs of emergency measures such as removing debris and repairing or replacing structures such as public buildings, roads, bridges, and public utilities.
3. **Hazard Mitigation.** FEMA funds mitigation measures to prevent or lessen the effects of a future disaster through the Hazard Mitigation Grant Program.[9]

Even if the President issues an emergency or major disaster declaration, not all persons or entities affected by a disaster are eligible for disaster assistance. FEMA officials determine the need for assistance from authorized categories after a declaration is issued and provides assistance only to those persons or entities determined to need the assistance.[10]

Fire Management Assistance Grant Declarations

The DRF is also used to fund the Fire Management Assistance Grant (FMAG) program. While the President has the sole authority to issue an emergency or major disaster declaration, the determination to issue a FMAG declaration can be rendered either by the President or FEMA.[11] In many instances they are issued by the FEMA Regional Director in coordination with FEMA headquarters. A FMAG declaration authorizes various forms of federal assistance, such as equipment, personnel, and grants to any state or local government for the control, management, and mitigation of any fire on public or private forest land or grassland that might become a major disaster.[12] FMAG declarations are relatively modest in cost when compared to emergency declarations and major disaster declarations. A review of declarations under the Fire Management Assistance Grant Program shows the most expensive year was 1998, in which 53 declarations were made, accounting for obligations of roughly $105 million.[13] By way of comparison, in 1998 $3.7 billion was obligated for major disaster declarations.

CURRENT FUNDING AND BUDGETING PRACTICES FOR THE DRF

The following section discusses how the DRF is funded including the budgeting and appropriation process. It also describes the Budget Control Act and the DRF Monthly Report.

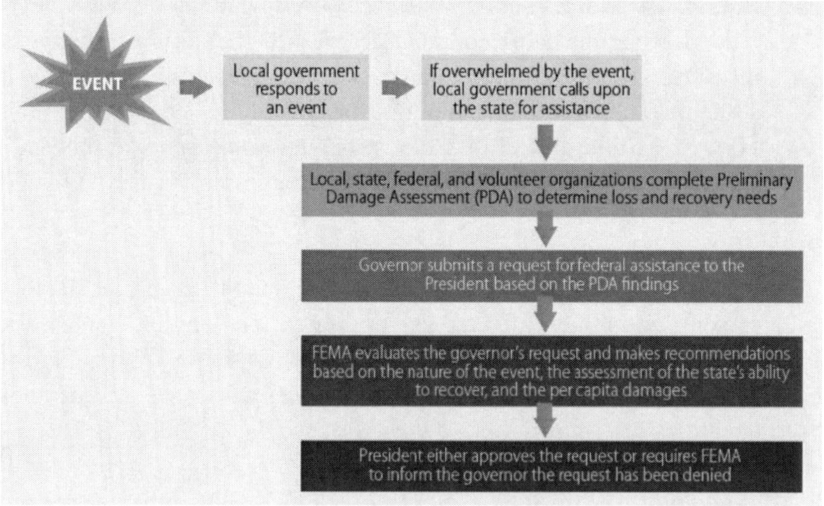

Source: CRS interpretation of statutes and federal regulations.

Note: Requests for emergency declarations can be made prior to an event. See Section 501(b) of the Stafford Act.

Figure 1. The Stafford Act Process for Declaring Emergencies and Major Disasters.

Annual Appropriations

Congress funds the DRF annually through regular appropriations, but unlike most appropriations which expire after a set period of time, the DRF is a "no-year" account. The funds for no-year accounts are available until expended—any remaining funds at the end of the fiscal year are carried over to the next fiscal year. One benefit of a no-year account is that the unobligated balance in the account can be used to pay for future disasters the next fiscal year. Another potential benefit of a no-year account is that the funds remain available during a government shutdown or an appropriation funding lapse.[14]

Supplemental Appropriations

In the past, funds in the DRF were often depleted before the end of the fiscal year due to disaster assistance needs. When the account nears depletion, Congress usually provides additional funding through one or more supplemental appropriations.[15] The need for additional funds is generally caused by a large-scale, major disaster such as Hurricanes Katrina or Sandy.[16] In recent years, however, the need for assistance has been increasingly tied to a string of incidents as opposed to a single, large event.

Table 1. Supplemental Funding for the DRF (in millions of dollars)

Fiscal Year	Number of Supplemental Appropriations	Amount
2013	1	$11,485
2012	1	$6,400
2011	0	0
2010	1	$5,100
2009	0	0
2008	3	$10,960
2007	1	$4,092
2006	2	$6,000
2005	3	$43,091
2004	2	$2,500
Total	**14**	**$89,628**

Source: Supplemental funding derived from CRS analysis of supplemental appropriations after disasters, FY2004-FY2013, with legislation obtained from the Legislative Information System at http://www.congress.gov/index.php.

Replenishing the DRF with supplemental appropriations has become common practice in the last ten years. As shown in **Table 1**, in some fiscal years Congress passed two or three additional appropriations to fund the DRF.

Budgeting the DRF

Funding the DRF at a level that meets disaster needs has been a persistent challenge for Congress. Historically chronic shortfalls in the DRF were attributed to two main factors: the previous decision not to budget for high-cost (over $500 million) catastrophic disasters in the annual appropriations process, and the unpredictability of the distribution of disaster events over time. **Figure 2** provides a list of major disasters that have cost more than $500 million from FY1996 to FY2013.

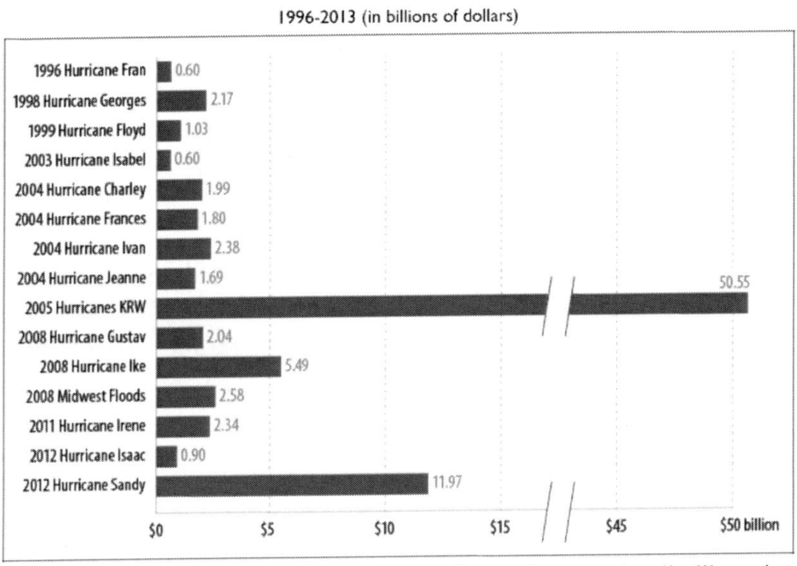

Source: Based on data obtained from http://www.fema.gov/media-library/assets/documents/28318?id=6292.

Notes: KRW denotes Katrina, Rita, and Wilma. Figures based on FEMA expenditures for disaster assistance. Figure 2 does not provide data on disaster funding data provided by other agencies. Spending on recovery continues for many of the listed events; these amounts represent a snapshot of disaster costs as of February, 2013.

Figure 2. Disasters Costing FEMA $500 Million or More 1996-2013.

Congressional concern over the number and amount of supplemental appropriations needed to fund disaster assistance has led to congressional debate concerning how the DRF should be budgeted, and whether the federal government is providing too much or too little assistance. These concerns are discussed more in depth in "The Debate over Supplemental Appropriations."

Budgeting for the DRF Prior to the BCA

The budgetary practice used to fund the DRF generally begins with the Administration's formulation of the annual budget request for the account. Prior to P.L. 112-25, the Budget Control Act of 2011(hereinafter the BCA) the following data points were used to determine budget requests for the DRF: (1) the available appropriation in the DRF; (2) the monthly average of "normal," non-catastrophic disaster costs paid from the DRF;[17] (3) the monthly average of catastrophic incident costs; and (4) the estimated monthly "recoveries" of unobligated funds.[18]

1. **Available Appropriation.** The available appropriation was a combination of prior-year funds that are carried over, the current fiscal year annual appropriation, and any supplemental appropriations.
2. **DRF Monthly Average.** The calculation for the DRF monthly average was based on a five-year rolling average of the cost of "normal," non-catastrophic disasters. Normal, non-catastrophic disasters were, and continue to be, defined by FEMA as incidents that cost less than $500 million. The rationale of excluding large events from the calculation is discussed later in this report.
3. **Monthly Cost Estimates for Catastrophic Events.** Estimates obtained from the field on pending (still open) disaster projects were routinely used in calculating monthly cost estimates.
4. **Estimated Recoveries.** Estimated recoveries represent the recovery of obligated funds that have not been used. This could include duplication of benefit funds as well as long-term projects for PA or mitigation that either were not finished, or were completed at a lower cost.

The end-of-fiscal-year projection was estimated by subtracting the cumulative DRF monthly averages and cost estimates for incidents from the available appropriation. Then, the cumulative recoveries were added back to

the available balance. The DRF end-of-fiscal-year estimate was then revised on a monthly basis taking into consideration the actual obligations that were recorded in lieu of the monthly estimates, and new estimates submitted for "open" incidents.

Based on the above methodology, the average annual budget request submitted for the DRF from FY2000 to FY2011 was roughly $2 billion. The average spend-out rate for the DRF over that same period was $350 million per month, or $4.2 billion a year.[19] It could be argued that without resources beyond the request and regular appropriation, the DRF would have faced a shortfall in its budget in an average operating year.

Budgeting for the DRF after the BCA

The Balanced Budget and Emergency Deficit Control Act of 1985 (hereinafter the BBEDCA), as amended by the BCA, limits—or caps—the budget authority available for discretionary programs each fiscal year through FY2021. The BBEDCA and the BCA establish discretionary spending caps but also provide adjustments—including an adjustment to disaster relief and an unlimited adjustment for emergency designations. The allowable adjustment to the discretionary cap for disaster relief is not solely for the DRF. This allowable adjustment can be changed and there are adjustments each fiscal year by no more than the average funding provided for disaster relief over the previous ten fiscal years—excluding the highest and lowest years—plus any amount by which the prior year's appropriation was below the maximum allowable cap adjustment for that year. The actual adjustment is determined during the appropriations process.

FEMA reports that the BCA necessitated the development of a new, two-part approach to accounting for disaster-related activity, with one approach for major disasters and another for all other DRF activity:

> Essentially, requests for DRF funding for FEMA's Stafford Act programs and disaster support activities fall into two categories: disaster relief cap adjustment and base/non-major disasters. Funding requested under the disaster relief cap adjustment is for major disasters declared pursuant to the Stafford Act and designated by the Congress as being for disaster relief pursuant to section 251(b)(2)(D) of the BBEDCA, as amended by the BCA. Funding requested under the base/non-major disasters category includes Emergencies, Pre-disaster Surge Support, Fire Management Assistance Grants and activities that are non-disaster specific, such as Disaster Readiness Support (DRS) activities (e.g., distribution centers, reservist training, etc.).[20]

The funding request for major disasters is based on FEMA's spending plans for all past declared major disasters. The non-catastrophic funding request is based on a revised approach that uses a 10-year average for non-catastrophic events. FEMA argued that using a 10-year average of costs as opposed to the previous use of a 5-year average of costs "provides a more accurate projection of non-catastrophic needs since it normalizes the effects of outlier years."[21]

As shown in **Table 2,** FEMA's new, two-part approach to accounting for disaster-related activity has resulted in higher Administration budget requests for DRF funding. For example, as mentioned previously, from FY2000 to FY2011, the average annual DRF budget request was $2 billion and the average annual spend-out rate was $4.2 billion. The Administration requested almost $6.1 billion for the DRF for FY2013 ($5.481 billion for the disaster relief cap adjustment and $608 million for base/non-major disasters) and nearly $6.2 billion in FY2014 ($5.626 billion for the disaster relief cap adjustment and $570.5 million for base/non-major disasters).[22] The enactment for those years were $7 billion and $6.2 billion respectively.

It appears that the allowable adjustment has allowed the DRF to be funded at historically high levels in recent years without supplementals (with the exception of the Sandy supplemental). However, this mechanism will expire, along with the caps, after FY2021. Some might question whether the problem of underfunding the DRF before the BCA will reoccur.

DRF Monthly Report

In addition to the BCA, Congress has passed additional legislation enabling it to exercise greater control over federal disaster assistance funding. For example, since Hurricane Katrina, a number of measures have been passed requiring FEMAto publish status reports on the DRF. P.L. 109-62 requires the Inspector General of DHS to conduct audits and investigations for Hurricane Katrina response and recovery activities and for the Secretary of DHS to provide (at a minimum) a weekly report to the Committees on Appropriations detailing the allocation and obligation of funds for Hurricane Katrina.[23] P.L. 109-90 requires FEMA to submit a report providing the details on allocations, obligations, and expenditures of funds from the DRF.[24]

Table 2. Requests, Appropriations, and Supplemental Appropriations to the DRF: FY1989-FY2014 (in millions of nominal dollars)

Fiscal Year	Administration Request	Enacted Appropriation	Supplemental	Total Appropriation
1989	$200	$100	$1,108	$1,208
1990	$270	$98	$1,150	$1,248
1991	$270	0	0	0
1992	$184	$185	$4,136	$4,321
1993	$292	$292	$2,000	$2,292
1994	$1,154	$292	$4,709	$5,001
1995	$320	$320	$3,275	$3,595
1996	$320	$222	$2,275	$2,497
1997	$320	$1,320	$3,300	$4,620
1998	$2,708	$320	$0	$320
1999	$2,566	$1,214	$1,130	$2,344
2000	$2,780	$2,780	0	$2,780
2001	$2,909	$1,600	$2,000a	$3,600
2002	$1,369	$2,164	$7,008	$9,172
2003	$1,843	$800	$1,426	$2,226
2004	$1,956	$1,789	$2,500	$4,289
2005	$2,151	$2,042	$43,091	$45,133
2006	$2,140	$1,770	$6,000	$7,770
2007	$1,941	$1,487	$4,256	$5,743
2008	$1,652	$1,324b	$10,960	$12,284
2009	$1,900	$1,278	0	$1,278
2010	$2,000	$1,600	$5,100	$6,700
2011	$1,950	$2,645	0	$2,645
2012	$1,800	$7,100	$6,400	$13,500
2013	$6,089	$7,007	$11,485	$18,492
2014	$5,626	$6,220	N/A	$6,220
Total	$46,710	$45,969	$123,084	$169,278

Source: CRS analysis of Administration budget documents and appropriations statutes.

Notes: Table 2 does not include rescissions or transfers unless they have been incorporated in appropriation acts. Table 2 also does not include appropriations made in the same act to accounts other than the DRF.

a. P.L. 107-38 appropriated $40 billion in response to the terrorist attacks of September 11, 2001. The legislation did not specify the amount to be allocated to the DRF, but required that not less than half must be allocated for disaster recovery and

assistance associated with the airliner crashes in New York, Virginia, and Pennsylvania. On September 21, 2001, President Bush notified Congress that $2 billion of the amount appropriated in P.L. 107-38 would be allocated to FEMA for disaster relief "in New York and other affected jurisdictions."

b. Does not include $2,900 million in FY2008 emergency supplemental funding for Disaster Relief enacted by P.L. 110-116.

The Post-Katrina Emergency Management Reform Act (P.L. 109-295, hereinafter PKEMRA) required that the report be submitted monthly and include greater detail. The report includes the status of the DRF (which consists of the obligations, allocations, and amounts that are undistributed or unallocated), information on DRF funding for Hurricanes Katrina, Rita, and Wilma, information on national flood insurance claims, funding information by state for unemployment, crisis counseling, housing assistance, public assistance, and individual assistance. The report also provides Mission Assignment obligations by agency. The requirements established by PKEMRA have since been reauthorized through various appropriations laws.[25]

Perhaps one of the most useful aspects of the DRF report to policymakers is the DRF Appropriations Summary (see **Figure 3** for an example). Since the passage of the BCA, the summary report now divides funding information into "Major Declarations" and "Base" categories (costs associated with every other activity funded through the DRF—such as support activities, emergency declarations, and FMAGs) and provides information on how much funding was received for the DRF through annual and supplemental appropriations. The summary page of the report provides information on allocations, commitments, and obligations.[26]

The report also provides information on funds recovered from DRF-funded projects. Recoveries are often funds from projects that were completed under budget and thus can be "recovered" because they remain unspent. In addition, in some cases DRF funds are disbursed before an insurance company pays for damages. In such cases the recipient must pay back the amount provided by the insurance company. These too are considered recoveries.

In the past there have been occasions when the DRF was near depletion. In an effort to keep the DRF solvent, FEMA implemented what is known as Immediate Needs Funding (INF) guidance.[27] INF allows FEMA to divert funds from long-term projects to focus on immediate, lifesaving response and recovery efforts. The INF guidance stayed in effect until the DRF was replenished. FEMA typically steps up its efforts in identifying recoveries when INF guidance is implemented because retrieving recoveries has been an effective stopgap measure when the account has been low. Over the last two or three years, however, FEMAhas become more adept as identifying recoveries in its day-to-day operations. As a consequence, FEMA may not be able to rely on as many recoveries in the future implementation of INF guidance. It is unclear how this might affect future efforts to keep the DRF solvent should it run low again.

ISSUES FOR CONGRESS

The DRF has been of Congressional interest for a number of reasons. These include the amount of funding appropriated to the DRF, the appropriateness and effectiveness of providing additional funding to the DRF through supplemental appropriations, and the use of policy mechanisms to reduce the amount of funding provided to states and localities for emergency and disaster assistance. One fundamental debate that has been of concern, particularly in the light of the national debt, is the federal role in providing disaster assistance.

The Debate over Supplemental Appropriations

There has been much debate over the supplemental appropriations for disaster assistance. Critics of the DRF budgetary process argued that the weaknesses in the methodology used to develop the budget request for the DRF led to the reliance on supplemental appropriations to fund major disasters. They argued that relying on supplemental appropriations for disaster assistance has the following drawbacks:

- supplemental appropriations for disasters often are designated as an emergency expenditure, which under congressional budgetary procedures can exceed discretionary spending limits designed to reduce the federal deficit—creating an opportunity for lawmakers to circumvent budgetary enforcement mechanisms by underfunding the DRF during the annual appropriations process to make room for other spending;
- supplemental appropriations for disasters often move through Congress on an expedited basis, limiting the amount of time available to assess actual disaster needs and scrutinize spending to ensure that

the spending is appropriately scaled, targeted, and that adequate safeguards are in place to address the potential for waste, fraud, and abuse. In addition, supplemental appropriations for disasters may result in unnecessarily high funding levels, as early damage estimates may overstate actual needs; and

- supplemental appropriations for disasters provide a vehicle for non-germane provisions in the legislation that may not pass on their own, or make the appropriation legislation contentious, thus slowing down the delivery of federal disaster assistance.

Selected Example: FY2014—Appropriation Summary

Department of Homeland Security - Federal Emergency Management Agency
Disaster Relief Fund Congressional Monthly Report
FY 2014 Activity
Through 01/01/14*

Dollars in Millions

Status of Obligations:	Major Declarations	Base	Total	Change Since Last Report
FY 2014 APPROPRIATION /CR (1)	6,076	577	6,653	0
FY 2013 Emergency Funding	0	0	0	0
FY 2014 Supplemental	0	0	0	0
Prior Years Deobligations (Recoveries)	113	139	251	65
Prior Years Carryover	6,682	1,810	8,492	0
Transfers (2)	0	-24	-24	0
FY 2014 APPROPRIATION AVAILABLE (3)	12,871	2,500	15,372	65
Less: TOTAL OBLIGATIONS	1,248	75	1,323	497
UNOBLIGATED BALANCE:	11,623	2,425	14,048	-432
Status of Commitments:				
Less: TOTAL COMMITMENTS	1,250	89	1,339	487
UNCOMMITTED BALANCE:	11,621	2,411	14,033	-422
Status of Allocations:				
TOTAL ALLOCATION	1,260	160	1,420	449
Less: TOTAL OBLIGATIONS	1,248	75	1,323	497
UNOBLIGATED ALLOCATION:	12	85	97	-48
Amount Undistributed/Unallocated	11,611	2,340	13,952	-384

NOTES:

* Data was pulled on 01/02/2014 due to the New Year's Day Holiday

(1) Pursuant to Sec. 110 of P.L. 113-46, OMB apportioned funding at a rate of operations equal to 29.32%, which has been made available for use thru 1-15-2014 ($1,944M)

(2) Transfer to DHS OIG

(3) Transfer authority into the Base is capped at 10% of Base Appropriation (currently estimated at $250M)

Source of financial information is the Integrated Financial Management Information System (IFMIS)

Source: Federal Emergency Management Agency (FEMA)'s monthly report FY2014, available at http://www.fema.gov/media-library-data/1389041387749-57cc1965d8712f40ae7ff b6153fc3c3c/FEMA+DRF+Through_1-1-2014.pdf.

Notes: As noted by the source: "Totals that have minor discrepancies are due to rounding." (See source for more details).

Figure 3. FEMA's Disaster Relief Fund Monthly Report.

Advocates of the use of supplemental disaster assistance would argue that:

- the timing and severity of disasters cannot be anticipated and appropriating a relatively large sum of funds through the regular, annual appropriations process may require Congress to reduce funding for other programs to pay for an unknown, and possibly non-existent, future event;
- the President is authorized to unilaterally determine when federal assistance is made available after a major disaster incident. Congress retains authority to control federal spending by voting on supplemental appropriations. In essence, the use of supplemental appropriations for disasters enables Congress to express its own preferences in disaster assistance;
- DRF balances may be subject to transfer or rescission, which may carry an additional negative consequence if a large disaster were to take place after the funds have been withdrawn. If this were to happen, another transfer or supplemental appropriation might be needed to address disaster needs; and
- supplemental appropriations for disasters can be sized according to the needs of the actual incident.

Proposals to Restructure Disaster Relief Budgetary Practices

Some proposals have been advanced to further reduce the need for supplemental appropriations through the restructuring of budgetary procedures.[28] Some of these options include the following:

Eliminate Any Unrelated Provisions in a Supplemental Bill

Supplemental appropriation bills may include a variety of funding and other provisions in a single bill that are unrelated to the incident. The pressing need for assistance may allow the passage of these unrelated measures. In addition, these elements could potentially prevent the passage of legislation that might pass if it were not attached to a supplemental appropriation. Prohibiting unrelated provisions may help reduce costs and eliminate controversial measures that could slow the passage of appropriations legislation.

Increase Annual Funding to the DRF

As mentioned earlier in this report, Congress may decide to increase the funding level of the DRF through annual appropriations. Doing so could eliminate, or at least greatly decrease, the need for supplemental funding. As noted earlier, an increase in the regular annual appropriations to the DRF appears to have been one result of the passage of the BCA.

Creation of a Rainy-Day Fund

A "rainy-day fund," also known as a reserve account, could be financed by cuts in other discretionary accounts, or through revenue-raising measures.[29] Spending from this fund would then only be allowed when needed for expensive disasters. Proponents of this policy option would likely argue that, in contrast to supplemental appropriations, which increase the federal deficit through borrowing funds, rainy-day accounts do not add to the federal deficit because they are funded through savings and/or revenue-raising measures. Furthermore, the balance for a rainy-day fund would increase during periods in which there are few or relatively small disasters.

Opponents might argue that a rainy-day fund is infeasible due to the high costs of catastrophic events. For example, Congress appropriated roughly $120 billion for Hurricane Katrina recovery efforts and $60 billion for Hurricane Sandy. Financing a fund capable of providing funds for events such as these through budget cuts and raising revenues would be difficult. Furthermore, once raised, a large fund might be subject to rescissions and transfers.

Creation of a Contingency Fund

A contingency fund based on a cost analysis of previous disasters could be created for use after a large disaster occurs. A contingency fund could be funded at a level sufficient for large disasters, while relatively routine disasters would still be funded through the DRF. Unlike a rainy-day fund—which pays for disasters through savings and revenue generating measures—a contingency fund would receive an annual appropriation. Funds from the contingency fund would only be disbursed under certain conditions or incidents. For example, there have been discussions concerning the addition of a new category of disaster declaration known as a "catastrophic declaration" for events characterized by extraordinary devastation.[30] Historic events that might qualify for a catastrophic declaration are the 1906 San Francisco earthquake and fire, the terrorist attacks of September 11, 2001, and Hurricane Katrina. A catastrophic declaration might be used for a nuclear bomb explosion, a tsunami hitting a highly populated area, or an immense and destructive earthquake, among others. The contingency fund could be the funding mechanism for catastrophic incidents.

On the other hand, some might question the feasibility of a contingency fund. For one, the appropriation amount for a contingency fund capable of paying for an incident such as Hurricane Katrina or Hurricane Sandy would be significant. In addition, large supplemental appropriations for disaster assistance have rarely been contentious because there is great a deal of sympathy toward disaster victims. It may be more difficult to pass a large appropriation in the absence of an incident—particularly in light of the federal deficit. Large-scale disasters are infrequent incidents. If left unused for long periods of time, the contingency fund may need to be adjusted for inflation to meet disaster needs. The contingency fund may also be subject to transfers if the fund is perceived as an unused resource.

Disaster Assistance Offsets

Some have proposed that supplemental funding should be "offset."[31] Appropriation legislation that is fully offset has no overall net cost in budget authority or outlays. Offsets can be achieved by cutting budget authority from one account and providing it to another account, or transferring budget authority from other programs. In recent years the debate over the use of offsets for disaster relief or assistance has intensified due to the growing size of the budget deficit and national debt.

As a result of recent congressional deliberations, legislative attempts have been made to offset the costs of disaster assistance. For example, Title VI of the House-reported version of H.R. 2017, the FY2012 Homeland Security Appropriations bill, would have provided an additional $1 billion of additional funding to the DRF by transferring resources from the Department of Energy. The provision reads as follows:

> Sec. 601. Effective on the date of the enactment of this Act, of the unobligated balances remaining available to the Department of Energy pursuant to section 129 of the Continuing Appropriations Resolution, 2009 (division A of P.L. 110-329), $500,000,000 is rescinded and $1,000,000,000 is hereby transferred to and merged with 'Department of Homeland Security—Federal Emergency Management Agency—Disaster Relief': Provided, That the amount transferred by this section is designated as an emergency pursuant to section 3(c)(1) of H.Res. 5 (112th Congress).[32]

Another example is the proposed amendment H.Amdt. 4 to Disaster Relief Appropriations Act, 2013[33] in the 113th Congress which would have provided an offset of the $17 billion in emergency funding to address the immediate needs for victims and communities affected by Hurricane Sandy. The offset would have been achieved by an across-the-board rescission of 1.63% to all discretionary appropriations for FY2013. The amendment was not adopted.

Proponents of offsets argue that they provide a mechanism to control spending and offset the costs of disaster assistance. Opponents argue that offsets politicize disaster assistance because the program selected for the offset may have been selected because it is politically unpopular rather than being based sound policy basis. They may also argue that the debate over the use of offsets will unnecessarily slow the delivery of needed assistance.

One potential argument against the sole reliance on offsets to limit federal spending on disaster assistance is that it fails to address the growing number of declarations issued each year. As the number of declarations increase over time, so too will their total cost. And as their total cost rises, more and more funding will be needed from other federal programs to fund offsets to subsidize disaster costs. In addition, a significant amount of funding would be needed to fully offset a large-scale disaster. Hurricanes Katrina and Sandy cost the federal government $120 billion and $60 billion respectively. As such, critics might argue that the sheer size of the offset might negatively impact other parts of the federal budget.

Proposals for Managing Declarations

Restructuring the budgetary process is one approach that may reduce the need for supplemental funding to pay for major disasters. Another approach would be to use various policy mechanisms to reduce the amount of funding the federal government provides for disaster assistance. These include reforming the declaration process, adjusting the federal share for assistance, and shifting some of the responsibility for paying for recovery to the state and/or the private sector.

Rationale for Keeping Declaration Policies and Disaster Assistance the Same

To many, providing relief to disaster victims is an essential role of the federal government. In their view, while the concern over costs is understandable given the potential impact of disaster assistance on the national budget, the number of declarations being issued each year and their associated costs are justified given the immediate and long-term needs created by incidents. They may further argue that providing assistance to disaster-stricken areas is both acceptable and needed to help a state and region's economy recover from an incident that it otherwise may not be able to recover from on its own. In addition, the costs of disasters should be expected given changes in severe weather patterns, as well as increases in population size and development.[34]

Limiting the Number of Major Disaster Declarations

Some may contend that too many major disasters are being declared and should be limited. Limiting the number of declarations could produce savings because declarations trigger funding for the DRF. The following sections review some policy mechanisms that could be employed to decrease the number of declarations that are being issued.

One option consists of preventing what may be perceived by some to be "marginal incidents" from triggering federal assistance. For example, some might question if a snowstorm or an ice storm are incidents that truly exceed the state's response capacity. They may further question whether these incidents are worthy of federal assistance. Potential methods for eliminating marginal incidents include changing the definitions of a major disaster in the Stafford Act, changing the per capita formula for determining whether a disaster is sufficiently large to warrant federal assistance, or the use of other indicators instead of, or in conjunction with, the per capita formula.

Changing the Definition of Major Disaster in the Stafford Act

Some argue that the Stafford Act has enhanced presidential declaration authority because the definition of a major disaster in Section 102(2) of the Stafford Act is, in their view, ill-defined.[35] Because of the expansive nature of this definition under the Stafford Act, they assert, there are not many restrictions on the types of major disasters for which the President may issue a declaration.[36] As noted previously, some would argue that snowstorms do not warrant major disaster declarations.

Operationally, changing the definition of a major disaster could also mean changing the definition(s) of the criteria used by FEMA to determine whether or not a major disaster is warranted.[37]

Changing the Per Capita Formula

One potential method of reducing the number of major disasters being declared is to increase the per capita amount used by FEMA to make major disaster recommendations to the President. A per capita formula based on damages caused by an incident is used by FEMA to make recommendations to the President concerning whether to issue a major disaster declaration. The current per capita amount used by FEMA to make recommendations is $1.32. Essentially, the estimated amount of public property damages caused by the incident is divided by the state's population. In general, if that amount exceeds $1.32 per person (in that state) FEMA will make a recommendation to the President that a major disaster declaration should be issued. The per capita amount of $1.32 could be increased (for example, by 10%) to reduce the number of incidents eligible for federal assistance.

The DHS Inspector General issued a report in May 2012 which noted that FEMA had been using a $1 per capita damage amount since 1986 to determine its recommendation (during its preliminary damage assessment process) to the President whether an event warranted federal assistance. The DHS Inspector General also explained that FEMA did not begin adjusting that number for inflation until 1999. The DHS Inspector General pointed out that if the inflation adjustment had been occurring over that 13-year period, from 1986 to 1999, fully 36% fewer disasters would have qualified for a presidential declaration based on that factor.[38]

The actual public announcement of factors used in considering a declaration did not become public until 1999. Until then, all of that information had been within the "pre-decisional" part of the process in the Executive Branch. That is not to say FEMA was not using the per capita amount in its considerations, only that the process was not widely known or understood as it presently is. As the DHS Inspector General noted, FEMA could have been raising that amount gradually since 1986. It is worth noting that when FEMA discussed such proposals (e.g., per capita figures gradually increasing) with Congress, the result was an amendment to the Stafford Act prohibiting the preclusion of a geographic area from receiving assistance solely by virtue of an arithmetic formula or sliding scale based on income or population.[39]

The Use of State Capacity Indicators

In 2001, the Government Accountability Office (GAO) issued a report on disaster declaration criteria. The GAO report was a comprehensive review of FEMA's declaration criteria factors. GAO recommended that FEMA "develop more objective and specific criteria to assess the capabilities of state and local governments to respond to a disaster" and "consider replacing the per capita measure of state capacity with a more sensitive measure, such as a state's total taxable resources."[40]

The state's Total Taxable Resources (TTR) were developed by the Department of the Treasury. GAO reported that TTR:

> is a better measure of state funding capacity in that it provides a more comprehensive measure of the resources that are potentially subject to state taxation. For example, TTR includes much of the business income that does not become part of the income flow to state residents, undistributed corporate profits, and rents and interest payments made by businesses to out-of-state stock owners. This more comprehensive indicator of state funding capacity is currently used to target federal aid to low-capacity states under the Substance Abuse and Mental Health Service Administration's block grant programs. In the case of FEMA's Public Assistance program, adjustments for TTR in setting the threshold for a disaster declaration would result in a more realistic estimate of a state's ability to respond to a disaster.[41]

It could be argued that the use of TTR would conflict with the prohibition against the use of arithmetic formulas established by Congress. However, just as FEMA's per capita measurement is one of several factors considered and not the "sole" determinant of a declaration, GAO stated that TTR would not violate Section 320 because TTR could also be used with other criteria such as those identified in regulations. Thus, some could contend that TTR could fill a similar role with perhaps more accuracy. TTR advocates also argue that it may also help reduce federal costs for disaster assistance by denying assistance to marginal incidents that could be otherwise handled by the state.

Expert Panels

Some have proposed the use of an independent expert panel to review gubernatorial requests for major disaster declarations.[42] Such panels would be comprised of individuals with specialized knowledge in certain subject areas, such as disasters, economics, and public health. The panel would take into account the severity of the incident as well as other factors that might indicate how well the state could respond to and recover from the incident. The panel would then make recommendations to the President whether the circumstances of the incident were worthy of federal assistance based on their assessment.

Some might argue that the use of an expert panel would make decisions about whether to provide assistance more objective. Others might argue that the use of a panel may slow down the declaration process and impede the provision of important assets and resources. It may be argued that the panel's recommendation would infringe on the President's authority to issue a declaration. On the other hand, it could also be argued that the President would retain the authority to issue a declaration despite the panel's recommendation.

Emergency Loans

Another potential method to reduce the number of declarations and the costs of federal disaster assistance would be to create incentives to dissuade states from requesting assistance. One method would be converting some, or all, federal assistance provided through emergency declarations into a loan program. For example, emergency declarations could be altered to provide up to a specified amount (for example, $5 billion) in low-interest recovery loans rather than or in addition to assistance grants.[43] Under this arrangement a state could elect to handle the incident without federal assistance rather than having to reimburse the federal government for recovery loans.

Similarly, another potential option would be expanding FEMA's Community Disaster Loan (CDL) program to include loans for disaster recovery.[44] Currently, the CDL program provides loan assistance to local governments that are having difficulty providing government services because of a loss in tax or other revenue following a disaster. The program assists local governments by offering federal loans to compensate for this temporary or permanent loss in local revenue.[45]

Proposals to Change Stafford Act Provisions

The following section discusses some potential changes to the Stafford Act that might limit the number of declarations being issued each year and thus reduce the demand for DRF funding.

Repeal Section 320

Section 320 of the Stafford Act restricts the use of an arithmetic or sliding scale to provide federal assistance. Repealing Section 320 would allow formulas that establish certain thresholds that states would have to meet to qualify for assistance. This might make declarations less discretionary and more predictable.

Amend Section 404

Section 404 of the Stafford Act[46] authorizes the President to contribute up to 15% of the cost of an incident toward mitigation measures that reduce the risk of future damage, loss of life, and suffering.[47] Section 404 could be amended to make mitigation assistance contingent on state codes being in place prior to an event. For example, states that have met certain mitigation standards could be eligible for a higher federal contribution for mitigation measures than states that do not meet the standards. The amendment may incentivize mitigation work on behalf of the state and possibly help reduce damages to the extent that a request for assistance is not needed, or the cost of the federal share may be lessened. The amendment could be set to take effect in five years, giving states time to act, or not.

Other Potential Amendments to the Stafford Act

Other amendments to the Stafford Act could either limit the number of declarations being issued, or the amount of assistance provided to the state by the federal government.

- The Stafford Act could be amended so that federal assistance would only be available for states with corollary programs (such as Public Assistance, Individual Assistance, and housing assistance). Establishing these programs at the state level may increase state capacity to handle some incidents without federal assistance. The amendment could be designed to take effect in five years, giving states time to act, or not.
- The Stafford Act could be amended to discontinue all assistance for snow removal unless directed by Congress. The amendment could be designed to take effect in five years to give states and localities an opportunity to increase snow removal budgets, or not.

Reducing the Amount of Assistance Provided through Declarations

Adjust the State Cost-Share

Under the Stafford Act, the federal share for assistance paid out of the DRF is 75%. In other words, state and local governments currently provide 25% of disaster costs on projects and grants to families and individuals with the federal government assuming 75% of all costs. However, it is useful to note that the 75% federal share can be increased if damages reach certain thresholds.[48] Some may contend, however, that efforts should be undertaken to reduce disaster costs by shifting more of the costs to the state and local levels by adjusting cost-shares.

Table 3 and **Figure 4** provides a hypothetical example of how adjusting cost-shares could potentially reduce federal funding for major disasters. The second column of **Table 3** represents federal obligations and supplemental appropriations that have been provided for selected incidents. The federal share for some of the incidents was greater than 75% (such as Hurricanes Katrina and Ike which, in some instances, received 100% of the federal share for certain programs). However, for the purposes of **Table 3** and **Figure 4** it is assumed that the federal share is 75%. If the federal share for disaster assistance was reduced by 10% (see third column), federal assistance would be reduced by a total of $21.5 billion. If the federal share was reduced by 25% (see column 4), federal assistance would be reduced by a total of $53.9 billion (nearly enough to offset the federal costs for Hurricane Sandy).

These savings are just for the large-scale disasters selected for this example. It is unclear how much savings could be garnered by reducing the federal share for all major disasters, but they could be significant.

There is no statutory limit on the number of people that can be helped following a disaster.[49] Similarly, when assessing damage to state and local infrastructure there is no cap on the amount of federal funds that can be expended to make the repairs or accomplish a replacement. The only limitation is that the damage must be to eligible facilities and that it is disaster-related damage.

Given that open-ended commitment by the federal government, some may argue that increasing the state share of 25% to a higher percentage would be warranted given the federal government's fiscal condition and competing priorities. Another option would be to make the cost-share arrangement not subject to administrative adjustment. Instead, the cost-share could only be adjusted through congressional action.

While some might contend that adjusting the federal cost-share would be an effective mechanism for reducing federal disaster costs, others might argue doing so would be burdensome to states and localities. For example,Arizona would have had to pay roughly $1.4 million dollars to meet a 50% matching requirement for the Wallow Fire in 2010.[50]

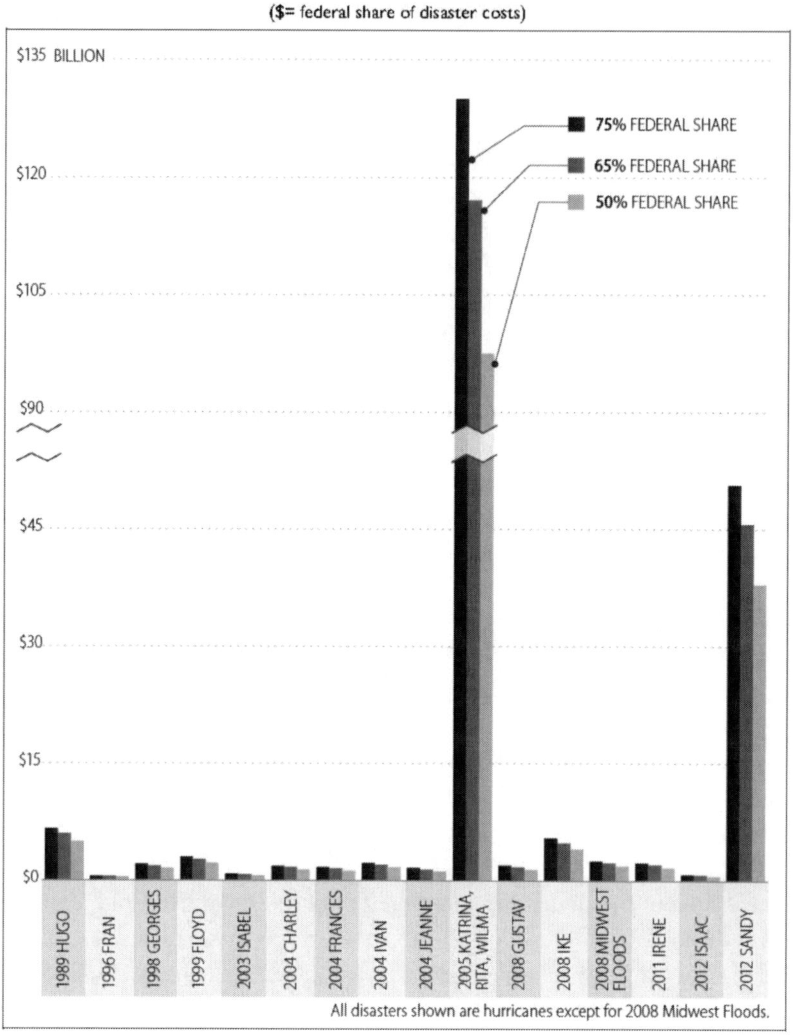

Source: Data derived from CRS interpretations supplemental appropriations, government studies, and reports.

Notes: A break in the funding data was used to display the funding information.

Figure 4. Examples of Select Potential Cost-Share Adjustments.

Disaster Loans

As mentioned previously, the assistance provided for emergency declarations could be provided through the form of loans. Similarly, some or all of the assistance provided to the state after a major disaster could be converted to low-interest or no-interest loans through the CDL program. For example, a state may receive the traditional 75% cost-share amount for an incident but be required to reimburse 25% of that total funding to the federal government. Loans for disaster recovery could also be incentivized. For instance, states that undertook certain pre-established preparedness and/or mitigation measures could qualify for a larger federal share or a lower interest rate.

Table 3. Examples of Select Potential Cost-Share Adjustments ($= federal share of disaster costs)

Incident	75% Federal Share	65% Federal Share	50% Federal Share
1989 Hurricane Hugo	$6,706,217,000	$6,035,595,300	$5,029,662,750
1996 Hurricane Fran	$604,802,216	$544,321,994	$453,601,662
1998 Hurricane Georges	$2,172,778,590	$1,955,500,731	$1,629,583,943
1999 Hurricane Floyd	$3,141,584,000	$2,827,425,600	$2,356,188,000
2003 Hurricane Isabel	$938,673,000	$844,805,700	$704,004,750
2004 Hurricane Charley	$1,985,931,385	$1,787,338,247	$1,489,448,539
2004 Hurricane Frances	$1,798,030,989	$1,618,227,890	$1,348,523,242
2004 Hurricane Ivan	$2,383,249,989	$2,144,924,990	$1,787,437,492
2004 Hurricane Jeanne	$1,691,106,423	$1,521,995,781	$1,268,329,817
2005 Hurricanes Katrina, Rita, Wilma	$129,999,411,000	$116,999,469,900	$97,499,558,250
2008 Hurricane Gustav	$2,043,546,976	$1,839,192,278	$1,532,660,232
2008 Hurricane Ike	$5,486,394,054	$4,937,754,649	$4,114,795,541
2008 Midwest Floods	$2,581,525,003	$2,323,372,503	$1,936,143,752
2011 Hurricane Irene	$2,340,660,813	$2,106,594,732	$1,755,495,610
2012 Hurricane Isaac	$900,343,795	$810,309,416	$675,257,846
2012 Hurricane Sandy	$50,700,000,000	$45,630,000,000	$38,025,000,000
Total	$215,474,255,233	$193,926,829,710	$161,605,691,425
Difference from 75% Share		**$21,547,425,523**	**$53,868,563,808**

Source: Data derived from CRS interpretations of supplemental appropriations, government studies, and reports.

Other Potential Policy Proposals

Oversight on Reporting

A 2006 Government Accountability Office (GAO) report indicated there was a need to improve the information in FEMA's weekly reports on the status of hurricane relief, and that OMB should take action to improve transparency and accountability regarding the status of hurricane-related funding.[51] Both OMB and FEMA agreed that these improvements were needed and would be forthcoming. Congress could authorize oversight mechanisms to investigate the extent to which FEMA has made such improvements. For example, section 203 of H.R. 5351 (introduced in the 109th Congress) would have required each state, local, tribal, and non-profit entity that received federal assistance funds in response to catastrophic events or other emergencies to report to the pertinent federal agency six months after the initial disbursement of resources. Furthermore, the legislation would have also required any agency that disbursed federal assistance funds to report to the Inspector General of the Department the purpose for which resources were provided, the amounts disbursed, allocated, and expended, and the status of reporting by agencies that received disbursements.

Concluding Policy Questions

Since the 1950s, the level of financial assistance given to states for disaster relief by the federal government has steadily increased. In light of stated concern with the federal deficit, the increased federal involvement has raised policymaking questions concerning how disaster relief should be equitably funded. Some of these questions include the following:

- The model for emergency and disaster response is built on the premise that emergencies and disasters are local. Requests for assistance from the next level of government are made only if the lower unit of government is overwhelmed. Some would argue that some incidents funded by the federal government do not meet this requirement. An example might be snow removal or repairs after minor flooding. Is the federal government funding emergency and major disaster declarations that do not meet the criterion of the states being overwhelmed before requesting assistance? Are states using federal funding for disaster relief to protect their own budgets?

- Should federal disaster relief be subject to thresholds and maximums? For example, an emergency or major disaster might not receive federal funding unless damage estimates reach a certain level. While the current system does use a per capita amount, that level could be increased. As another example, the total amount of federal relief for an event could be capped at a certain amount. After this level has been reached, the state would then be responsible to pay for the rest of recovery.
- Should the state's fiscal capability factor into disaster relief? In 1986, FEMA proposed measures to reduce the amount the agency contributed toward disaster relief. One of the proposals argued that funding allocations should be made according to each state's ability to fund its own disaster relief. The determination would be based on a comparison of the state's per capita income with the national per capita income.[52] The calculation would then be used to create a sliding scale for assistance. States capable of funding their own disaster relief would receive limited or no assistance. In contrast, struggling communities would be eligible to receive more federal assistance.[53]
- Is federal assistance to states and localities unintentionally creating a disincentive for states and localities to prepare for emergencies and major disasters? Some may argue that federal funding for disaster relief through regular annual appropriations has become entrenched to the point that it has contributed to unintended consequences. For example, it has been argued that some states do not properly fund mitigation measures because there is a presumption that federal funding is virtually guaranteed should an emergency or major disaster occur.[54] Those advocating this position could arguably point out that federal involvement in disaster relief will continue to increase and that in order to be fiscally responsible, changes should be made in the way in which disaster relief is funded. Others may claim the function of the federal government is to help states in their time of crisis. From this perspective, withholding or limiting the amount of funding a state could receive for an incident would be neglectful of that state's needs.
- As mentioned earlier, funding the DRF at a higher level through annual appropriations may give some the perception that the funds are not being used and could therefore be subject to rescissions or transfers. If larger appropriations for the DRF witnessed since the enactment of the BCA continues, will there be a temptation to transfer unobligated funds to other disaster assistance programs not authorized

by the Stafford Act? For example, could the funds from the DRF be used to fund drought relief programs provided by the United States Department of Agriculture, or to fund fire assistance provided by the Department of the Interior?[55] Would subsidizing other assistance programs negate the benefit of having a larger appropriation?

These and other questions may be raised should Congress elect to debate the past and future funding of disaster relief.

End Notes

[1] Congress also appropriates disaster funds to other accounts administered by other federal agencies pursuant to federal statutes that authorize specific types of disaster relief.

[2] Certain federal entities, such as the Department of Transportation, fund emergency and disaster assistance through their own budgets.

[3] 42 USC 5121 et seq. For information on the declaration process, see CRS Report RL34146, *FEMA's Disaster Declaration Process: A Primer*, by Francis X. McCarthy, and CRS Report R42702, *Stafford Act Declarations 1953-2011: Trends and Analyses, and Implications for Congress*, by Bruce R. Lindsay and Francis X. McCarthy. Note that some activities may be funded by the DRF without a presidential declaration. For example, the Disaster Readiness and Support Account pays for FEMA's phone centers, finance centers, and housing inspectors. Through this account certain recovery elements are already in place when the President issues a declaration.

[4] P.L. 85-536, Section 7(b) 72 Stat. 387, as amended.

[5] For more information on the SBA Disaster Loan Program see CRS Report R41309, *The SBA Disaster Loan Program: Overview and Possible Issues for Congress*, by Bruce R. Lindsay.

[6] For more information on agriculture disaster assistance see CRS Report RS21212, *Agricultural Disaster Assistance*, by Dennis A. Shields.

[7] 42 U.S.C. § 5122. Emergencies are defined as "any occasion or instance for which, in the determination of the President, Federal assistance is needed to supplement state and local efforts and capabilities to save lives and to protect property and public health and safety, or to lessen or avert the threat of a catastrophe in any part of the United States."

[8] Ibid. Major disasters are defined as any natural catastrophe (including any hurricane, tornado, storm, high water, wind driven water, tidal wave, tsunami, earthquake, volcanic eruption, landslide, mudslide, snowstorm, or drought), or, regardless of cause, any fire, flood, or explosion, in any part of the United States, which in the determination of the President causes damage of sufficient severity and magnitude to warrant major disaster assistance under this act to supplement the efforts and available resources of states, local governments, and disaster relief organizations in alleviating the damage, loss, hardship, or suffering caused thereby.

[9] A structure does not have to be damaged to be eligible for mitigation. Hazard Mitigation Grants are available as block grant funds and do not have to be used in the designated disaster area—they can be used anywhere in the state. For more information, see CRS Report R40471, *FEMA's Hazard Mitigation Grant Program: Overview and Issues*, by Natalie Keegan.

[10] For example, under 44 C.F.R. § 206.101(2) a household that has access to alternative housing would be ineligible for housing assistance.

[11] 44 C.F.R. 204.24.

[12] P.L. 93-288, 42 U.S.C. §5187(a).

[13] DHS/FEMA, Calendar Year Summary of Obligations, 1988-2010.

[14] For more information on appropriation funding lapses, see CRS Report RS20348, *Federal Funding Gaps: A Brief Overview*, by Jessica Tollestrup.

[15] For further analysis on emergency supplemental appropriations see CRS Report R40708, *Disaster Relief Funding and Supplemental Appropriations for Disaster Relief*, by Bruce R. Lindsay and Justin Murray.

[16] For more information on Hurricane Katrina funding see CRS Report R43139, *Federal Disaster Assistance after Hurricanes Katrina, Rita, Wilma, Gustav, and Ike*, coordinated by Bruce R. Lindsay and Jared C. Nagel. For more information on Hurricane Sandy funding see CRS Report R42869, *FY2013 Supplemental Funding for Disaster Relief*, coordinated by William L. Painter and Jared T. Brown.

[17] Normal non-catastrophic disasters are declared incidents that cost less than $500 million. Disasters costing over $500 million are considered outliers and are removed from the calculation.

[18] These may occur, for example, when funds remain unspent after a project is completed for less than the estimated cost or when a project for which funds have been obligated changes its scope and certain budgeted costs become ineligible.

[19] The spend-out rate refers to the amount of money paid out of the account for a given period of time.

[20] U.S. Department of Homeland Security, "Federal Emergency Management Agency Disaster Relief Fund: FY2013 Congressional Justification," p. 5, at http://www.fema.gov/pdf/about/budget/ 11f_fema_disaster_relief_fund_dhs_fy13_cj.pdf.

[21] Ibid., pp. 5-6.

[22] Ibid., p. 5; and U.S. Department of Homeland Security, "U.S. Department of Homeland Security Annual Performance Report Fiscal Years 2012–2014," p. 11, at http://www.dhs.gov/sites/default/files/publications/MGMT/DHS-%20Annual%20 Performance%20Report%20and%20Congressional-Budget-Justification-FY2014.pdf.

[23] P.L. 109-62, 119 Stat. 1991, the Second Emergency Supplemental Appropriations Act to Meet Immediate Needs Arising From the Consequences of Hurricane Katrina, 2005.

[24] P.L. 109-90, 119 Stat. 2090, the Department of Homeland Security Appropriations Act, 2006.

[25] These include P.L. 110-329, P.L. 111-83, P.L. 112-74, and P.L. 113-6. The DRF report is available online and can be located at http://www.fema.gov/disaster-relief-fund.

[26] For the purposes of the DRF report, allocations are funds for a particular program. Commitments are a reservation of allotted funds in anticipation of their obligation. Obligations are payments made immediately, or in the near future.

[27] For more information on FEMA's INF guidance see http://www.fema.gov/media-library-data/20130726-1847-25045-6072/9570.7_immediate_needs_funding_sop.pdf.

[28] U.S. Congress, Senate Bipartisan Task Force on Funding Disaster Relief, *Federal Disaster Assistance*, Report of the Senate Task Force on Funding Disaster Relief, 104th Cong., 1st sess., March 15, 1995, S.Doc. 104-4 (Washington: GPO, 1995), p. 89.

[29] Ibid, p. For more information about the funding of reserve or contingency funds see CRS Report R41718, *Federal Deposit Insurance for Banks and Credit Unions*, by Darryl E. Getter; CRS Report R42850, *The National Flood Insurance Program: Status and Remaining Issues for Congress*, by Rawle O. King; and CRS Report R43391, *Independence of Federal Financial Regulators*, by Henry B. Hogue, Marc Labonte, and Baird Webel.

[30] For more information on catastrophic declaration proposals see CRS Report R41884, *Considerations for a Catastrophic Declaration: Issues and Analysis*, by Bruce R. Lindsay and Francis X. McCarthy.

[31] For more information on offsets and supplemental appropriations see CRS Report R42458, *Offsets, Supplemental Appropriations, and the Disaster Relief Fund: FY1990-FY2013*, by William L. Painter.

[32] This section was added in full committee markup of the legislation. For a more in-depth discussion of procedural considerations for offsetting amendments, see CRS Report RL31055, *House Offset Amendments to Appropriations Bills: Procedural Considerations*, by Jessica Tollestrup.

[33] P.L. 113-2.

[34] For information on disaster trends see CRS Report R42702, *Stafford Act Declarations 1953-2011: Trends and Analyses, and Implications for Congress*, by Bruce R. Lindsay and Francis X. McCarthy.

[35] P.L. 93-288, 42 U.S.C. §5122.

[36] Richard T. Sylves, *Disaster Policy and Politics: Emergency Management and Homeland Security* (Washington, DC: CQ Press, 2008), p. 79.

[37] These criteria include a per capita formula and state capacity indicators, among others.

[38] Department of Homeland Security, Office of Inspector General, *Opportunities to Improve FEMA's Public Assistance Preliminary Damage Assessment Process*, pp. 5-7.

[39] 42 U.S.C. 5163. P.L. 100-707, the Major Disaster Relief and Emergency Assistance Amendments of 1987.

[40] U.S. Government Accountability Office, *Disaster Assistance; Improvement Needed in Disaster Declaration Criteria and Eligibility Assurance Procedures*, GAO-01-837, August 2001, p. 21.

[41] Ibid, p. 11.

[42] For example, S. 1630, the Disaster Recovery Act of 2011, which was introduced on September 23, 2011, would have amended the Stafford Act to authorize the President to declare a catastrophic incident if a recommendation was issued by an independent panel of experts.

[43] FEMA must notify Congress when an incident exceeds $5 million. Additional spending for the incident, if needed, must be authorized by Congress.

[44] For more information on the Community Disaster Loan program see CRS Report R42527, *FEMA's Community Disaster Loan Program: History, Analysis, and Issues for Congress*, by Jared T. Brown.

[45] The CDL program is funded through the Disaster Assistance Direct Loan Program account, rather than the DRF.

[46] 42 U.S.C. §5170c.

[47] For more information on hazard mitigation grants see CRS Report R40471, *FEMA's Hazard Mitigation Grant Program: Overview and Issues*, by Natalie Keegan.

[48] For additional discussion on this topic see CRS Report R41101, *FEMA Disaster Cost-Shares: Evolution and Analysis*, by Francis X. McCarthy.

[49] There is however, a limit on how much any one household can receive ($32,200 at the time of this report).

[50] Based on estimation of FEMA obligations for the incident.

[51] U.S. Government Accountability Office, *Disaster Relief: Governmentwide Framework Needed to Collect and Consolidate Information to Report on Billions in Federal Funding for the 2005 Gulf Coast Hurricanes*, GAO-06-834, September 2006.

[52] U.S. Congress, House Committee on Public Works and Transportation, Subcommittee on Investigations and Oversight, *The Federal Emergency Management Agency's Proposed Disaster Relief Regulations (Budget Driven Rulemaking)*, committee print, 100th Cong., 1st sess., August 1987, 75-963 (Washington: GPO, 1987), pp. 4-5.

[53] Under current law (42 U.S.C. 5163) areas cannot be precluded from receiving assistance *solely* on the basis of a sliding scale or formula. Congress amended the Stafford Act with Section 320 as a response to the 1986 proposed regulation.

[54] See James F. Miskel, *Disaster Response and Homeland Security: What Works, What Doesn't* (Westport, CT: Praeger Security International, 2006), or Rutherford H. Platt, *Disasters and Democracy: The Politics of Extreme Natural Events* (Washington, DC: Island Press, 1999) for arguments against increased federal involvement in disaster assistance.

[55] For more information on federal funding for wildfires see CRS Report RL33990, *Federal Funding for Wildfire Control and Management*, by Kelsi Bracmort. For more information on drought assistance see CRS Report RL34580, *Drought in the United States: Causes and Issues for Congress*, by Peter Folger, Betsy A. Cody, and Nicole T. Carter.

In: FEMA's Disaster Relief Fund
Editor: Angelica Conner
ISBN: 978-1-63321-630-3

Chapter 2

DISASTER RELIEF FUNDING AND SUPPLEMENTAL APPROPRIATIONS FOR DISASTER RELIEF*

Bruce R. Lindsay and Justin Murray

SUMMARY

When a state is overwhelmed by an emergency or disaster, the governor may request assistance from the federal government. Federal assistance is contingent on whether the President issues an emergency or major disaster declaration. Once the declaration has been issued the Federal Emergency Management Agency (FEMA) provides disaster relief through the use of the Disaster Relief Fund (DRF), which is the source of funding for the Robert T. Stafford Emergency Relief and Disaster Assistance Act response and recovery programs. Congress appropriates money to the DRF to ensure that funding for disaster relief is available to help individuals and communities stricken by emergencies and major disasters (in addition, Congress appropriates disaster funds to other accounts administered by other federal agencies pursuant to federal statutes that authorize specific types of disaster relief).

* This is an edited, reformatted and augmented version of a Congressional Research Service publication, CRS Report for Congress R40708, dated August 5, 2013.

Historically, the DRF is generally funded at a level that is sufficient for what are known as "normal" disasters. These are incidents for which DRF outlays are less than $500 million. When a large disaster occurs, funding for the DRF may be augmented through emergency supplemental appropriations. A supplemental appropriation generally provides additional budget authority during the current fiscal year to (1) finance activities not provided for in the regular appropriation; or (2) provide funds when the regular appropriation is deemed insufficient.

This methodology used to budget the DRF appears to have been altered since the passage of the Budget Control Act (P.L. 112-25). The Budget Control Act includes a series of provisions that directed the Office of Management and Budget (OMB) to annually calculate an "allowable adjustment" for disaster relief to the BCA's discretionary spending caps. That adjustment, if used, would make additional budget authority available for the federal costs incurred by major disasters declared under the Stafford Act beyond what is allowed in the regular discretionary budget allocation. The OMB calculation may have provided a mechanism that encourages a larger regular appropriation to the DRF. It is possible that larger DRF appropriations may reduce the need for supplemental appropriations.

Budgeting for disaster relief has been the subject of a great deal of debate. Some argue that more money should be appropriated in FEMA's DRF account in annual appropriations, while others maintain that augmenting the DRF through supplemental appropriations is preferable because it allows Congress to react directly to a particular situation. Others may argue that emergency supplemental appropriations are preferable for fiscal management reasons because an appropriation is not requested unless there is a real need for supplemental funding. Another argument is to revamp the budgetary process to fund disaster relief.

This report describes the various components of the DRF, including (1) what authorities have shaped it over the years; (2) how FEMA determines the amount of the appropriation requested to Congress (pertaining to the DRF); and (3) how emergency supplemental appropriations are requested. Information is also provided on funds appropriated in supplemental appropriations legislation to agencies other than the Department of Homeland Security (DHS). Aspects of debate concerning how disaster relief is budgeted are also highlighted and examined, and alternative budgetary options are summarized.

DISASTER RELIEF FUND

The Disaster Relief Fund (DRF), sometimes referred to as the President's Disaster Relief Fund, is managed by the Federal Emergency Management Agency (FEMA). The DRF is the main account used to fund a wide variety of programs that provide grants and other support to assist state and local governments and certain nonprofit entities during disaster recovery. In most cases, funding from the DRF is released after the President has issued a declaration pursuant to the Robert T. Stafford Relief and Emergency Assistance Act (Stafford Act).[1] There are, however, some activities that may be funded by the DRF without a presidential declaration, mainly those supported by the Disaster Readiness and Support Account. The Disaster Readiness and Support Account pays for FEMA's phone centers, finance centers, and housing inspectors. Through this account certain recovery elements are already in place when the President issues a declaration.

The declaration process has the following general structure:

- the local government responds to an event;
- if the event overwhelms the local government, the state is called upon for assistance;
- a Preliminary Damage Assessment (PDA) is completed by local, state, federal, and volunteer organizations to determine loss and recovery needs;
- based on the PDA findings, a declaration request is submitted to the President byt he governor;[2]
- FEMA evaluates the governor's request and recommends action to the White House, based on the nature of the event and an assessment of the state's ability to recover; and
- the President may either approve the request or require FEMA to inform the governor the request has been denied.[3]

A President's declaration triggers the allocation of funds from the DRF, and the funding may be distributed from any one, or any combination, of three categories of disaster aid:

1. **Individual Assistance.** Individual Assistance (IA) includes disaster housing for displaced individuals, grants for needs not covered by insurance, crisis counseling, and disaster-related unemployment assistance.

2. **Public Assistance.** Public assistance (PA), which is FEMA's largest funded program, helps communities absorb the costs of emergency measures such as removing debris and repairing or replacing structures such as public buildings, roads, bridges, and utilities.
3. **Hazard Mitigation.** FEMA funds mitigation measures to prevent or lessen the effects of a future disaster through the Hazard Mitigation Grant Program.[4]

Even if the President issues an emergency or major disaster declaration, not all persons or entities affected by a disaster are eligible for disaster assistance. FEMA officials determine the need for assistance after a declaration is issued. Aid is provided only to those persons or entities determined by FEMA to need the assistance. An example of ineligibility would be a household that has access to alternative housing.[5]

PAST AND PRESENT AUTHORITIES RELATED TO THE DISASTER RELIEF FUND

The Stafford Act authorizes appropriations to carry out disaster relief activities, but does not explicitly designate the DRF as the account for such funds. Rather, the DRF is the product of legislation and federal policies that can be traced to the post-World War II era. Prior to that time, disaster response activities were funded primarily through local efforts and voluntary groups. In cases where the federal government did offer assistance, the needs of disaster victims and affected communities were funded on an as-needed basis through appropriations that were then allocated, pursuant to the legislation, by executive branch administrators and, ultimately, the President.[6]

Arguably, the forerunners to the DRF were appropriations known as "emergency funds for the President." For example, in a 1948 appropriation, Congress allocated $500,000 to "supplement the efforts and available resources of state and local governments or other agencies, whenever ... any flood, fire, hurricane, earthquake, or other catastrophe in any part of the United States is of sufficient severity and magnitude to warrant emergency assistance by the federal government."[7] In subsequent years Congress continued to appropriate funds for disaster relief and eventually a series of authorizations were provided in Section 606 of the Disaster Relief Act of 1974 as amended.[8] These authorizations provided funding for disaster-related activities until a specified date. In 1988 Congress repealed Section 606 of the Disaster Relief Act ending extensions of authorizations.[9] Since then funds have been appropriated, though not specifically authorized.

Public Laws Influencing the Administration of Disaster Relief

The approach to disaster relief changed from 1950 to 1979, transitioning from a largely uncoordinated and decentralized system of relief funding to one dominated by the federal government. The transition was due in part to an increased awareness of the complex responsibilities of federal and nonfederal entities before and after a disaster, and to the efforts of policymakers and administrators to address a variety of concerns—most notably civil defense.[10]

After World War II, concern over the possible use of atomic weapons and growing hostility between the United States and the Soviet Union gave rise to the Cold War. As a consequence, disaster management in the United States was organized around two tracks: (1) the threat of a nuclear war and (2) natural disasters. Several landmark federal disaster laws and policies originate from attempts by lawmakers during this era to prepare the civilian population for a potential atomic attack and provide aid after a natural disaster.[11] The most notable of these laws were the Civil Defense Act of 1950[12] and the Federal Disaster Relief Act of 1950.[13] These laws set into motion federal-to-state assistance, prompting the need for an account to fund disaster and emergency activities.

Once a framework of federal to state disaster assistance had been constructed in the 1950 statute, the process of administering disaster relief was further shaped by the Disaster Relief Acts of 1966[14] and 1974,[15] and the Robert T. Stafford Disaster Relief and Emergency Assistance Act of 1988.[16] Prompted by large disasters such as Hurricanes Betsy in 1965 and Agnes in 1972, these laws significantly increased federal involvement in emergency management, which in turn created the need for a funding mechanism to pay for emergency and disaster activities. Summaries of the statutes follow.

The Federal Disaster Relief Act of 1950 (P.L. 81-875)

The Federal Disaster Relief Act established much of the framework through which disaster policy is carried out in the United States. The Federal Disaster Relief Act provided an orderly, ongoing continued means of federal assistance to states and localities to alleviate suffering and damage that resulted from major disasters. Prior to the act, congressional action after each individual incident was needed to provide federal aid. Once in place, the law authorized the President to make the decision to provide federal aid without the consent of Congress. The act also put in place the standard process in which the governors could ask the President for federal disaster assistance for their respective states.[17]

Disaster Relief Act of 1966 (P.L. 89-769)

To some, the Disaster Act of 1950 appeared to effectively handle routine disasters, but did not adequately address large-scale, catastrophic disasters. In response to this need, Congress enacted the Disaster Relief Act of 1966. Some of the measures in the act included the authorization of federal agencies to provide disaster loans below market rates, and the extension of aid to unincorporated areas.

Disaster Relief Act of 1974 (P.L. 93-288)

The Disaster Relief Act of 1974 contained several precedent-setting features. The act created the first program to provide direct assistance to individuals and households following a disaster. It instituted the Individual and Family Grant (IFG) program, which supplied 75% of the funding for state-administered programs that provided money to purchase clothing, furniture, and essential needs following a disaster. The act also formalized efforts to mitigate disasters, as opposed to merely responding to them. Additionally, the act stressed a multi-hazard approach wherein government officials would prepare and respond to all types of disasters, rather than maintaining separate capacities for different types of hazards.[18]

The Robert T. Stafford Disaster Relief and Emergency Assistance Act (P.L. 93-288)

Today, the Robert T. Stafford Disaster Relief and Emergency Assistance Act constitutes the statutory authority for most federal disaster response and recovery activities especially as they pertain to FEMA and FEMA programs. The act authorizes the President to issue major disaster, emergency, and fire management declarations at the request of the states and tribal governments.[19] The declarations enable federal agencies to provide assistance to states overwhelmed by disasters. Stafford Act disaster assistance is provided through funds appropriated to the DRF. The funds may be used by states, localities, and certain non-profit organizations for activities such as providing mass care, restoring damaged or destroyed facilities, clearing debris, aiding individuals and families with uninsured needs, and mitigating the impact of future disasters.[20]

The history of federal disaster relief legislation demonstrates a steady expansion of federal aid since the 1950s. As the federal role in emergencies and major disasters expanded, so too have the costs. The costs associated with disaster relief have led some to contemplate changing how the federal government funds these events. Some of these proposals are discussed later in this report.

Post-Katrina Emergency Management Reform Act of 2006 (P.L. 109-295)

In the aftermath of Hurricane Katrina, Congress passed the Post-Katrina Emergency Management Reform Act[21] to address emergency preparedness and response shortcomings identified in the reports published by congressional committees and the White House. Based on those reports and oversight hearings on many aspects of FEMA's performance during the hurricane season of 2005, Congress expanded the authority of the FEMA administrator, authorized accelerated federal aid, and raised some ceilings on federal assistance, among other changes. In addition, Congress mandated that, for FY2007, FEMA had to submit to Congress a monthly report on the DRF detailing allocations, obligations, and expenditures for Hurricanes Katrina, Rita, and Wilma.[22] The reports had to include information on national flood insurance claims, manufactured housing data, obligations, allocations for housing assistance, public assistance, individual assistance and expenditures by state for unemployment.[23]

How the DRF Is Funded

FEMA receives appropriations for disaster relief through annual appropriations. Each fiscal year, FEMA and the Office of Management and Budget (OMB) submit a request to the President for the amount of funding the two agencies determine the DRF should receive. The President then submits an Administration request to Congress. The President may, or may not, use the amount suggested by FEMA and the OMB.

The methodology previously used to determine the budget recommendation incorporated four data points. The data points included (1) the available appropriation, (2) the DRF monthly average (the amount in the

DRF), (3) the monthly cost estimates for catastrophic events, and (4) the estimated monthly recoveries of unobligated funds.[24]

1. **Available Appropriation.** The available appropriation was a combination of prior-year funds that are carried over, the current fiscal year appropriation, and any supplemental appropriation funding.[25]
2. **DRF Monthly Average.** The calculation for the DRF monthly average was based on a five-year rolling average "normal" disaster's costs. Normal disasters were considered to be incidents that cost less than $500 million. The rationale of excluding large events from the calculation is discussed later in this report.
3. **Monthly Cost Estimates for Catastrophic Events.** Estimates obtained from the field on pending (still open) disaster projects were routinely used in calculating monthly cost estimates.
4. **Estimated Recoveries.** Estimated recoveries represent the recovery of obligated funds that have not been used. This could include duplication of benefit funds[26] as well as long-term projects for PA or mitigation that either were not finished, or were completed at a lower cost.

The end-of-fiscal-year projection was estimated by subtracting the cumulative DRF monthly averages and cost estimates for incidents from the available appropriation. Then, the cumulative recoveries were added. The DRF end-of-fiscal-year estimate was then revised on a monthly basis taking into consideration the actual obligations that were recorded in lieu of the monthly estimates, and new estimates submitted for "open" incidents.[27]

The methodology described above may have led to a series of budget requests that were too low to meet disaster assistance needs solely through regular appropriations. In particular, the elimination of incidents of $500 million or more from the calculation may have led to a lower budget projection. The passage of the Budget Control Act (P.L. 112-25, hereinafter the BCA) may have altered how the DRF is now being budgeted. The BCA includes a series of provisions that directed the Office of Management and Budget (OMB) to annually calculate an "allowable adjustment" for disaster relief to the BCA's discretionary spending caps. That adjustment, if used, would make additional budget authority available for the federal costs incurred by major disasters declared under the Stafford Act beyond what is allowed in the regular discretionary budget allocation. The OMB calculation may have provided a mechanism that encourages a larger regular appropriation to the DRF. It is possible that larger DRF appropriations may reduce the need for supplemental appropriations. In addition, in the past there may have been more pressure in previous years to quickly pass a supplemental appropriation to meet the needs of the disaster. Higher appropriations for the DRF through regular appropriations may provide Congress with more time to debate disaster needs and target disaster assistance needs more efficiently when large scale disasters occur. For example, when Hurricane Sandy struck the northeastern United States in October 2012, the DRF had roughly $7 billion to meet the immediate demands of the hurricane. In previous years, when a large scale disaster occurred, the DRF balance was generally low due to smaller regular appropriations to the account.

A comparison between Administration requests and annual appropriations is presented in **Table 1.**

THE DEBATE OVER EMERGENCY SUPPLEMENTAL APPROPRIATIONS

Congress has traditionally appropriated funds to maintain the DRF at a certain level, and then provided additional financing for assistance through supplemental appropriations following a specific large disaster. Currently, the DRF funds disaster relief for emergencies and major disasters that cost $500 million or less. Major disasters costing more than $500 million are generally funded with emergency supplemental appropriations. A supplemental appropriation provides additional budget authority during the current fiscal year either to finance activities not provided for in the regular appropriation, or to provide funds when the regular appropriation is deemed insufficient.

Catastrophic incidents may deplete the DRF. In such cases, the President may submit a request to Congress for an emergency supplemental appropriation. Historically, FEMA is the second-largest recipient of supplemental appropriations.[28] Most emergency supplemental appropriations originate with the Administration. Some emergency supplemental appropriations, however, have been initiated by Congress. For example, in 2007, Congress waived the cost share provisions for Hurricanes Katrina and Rita, which prompted a supplemental appropriation (P.L. 110-28). Another example of an emergency supplemental appropriation originating in Congress was in 2008, when Congress added DRF supplemental funding (P.L. 110-329) to the FY2009 Department of Homeland Security Appropriations bill on its own initiative in response to the Midwest Flooding and Hurricanes Ike and Gustav. The amount of funding for the supplemental appropriation was based on estimates provided to Congress by FEMA.

The current process of funding disasters has been a subject of debate at times because of concern over high levels of federal spending and the inability to institute fiscal planning mechanisms. Some of the debate includes discussions of options for lowering the budget deficit through budgetary enforcement procedures in Congress. Others argue supplemental appropriations are used to make budgets appear to forecast a lower deficit. OMB indicated a desire to break from the current practice of funding disaster relief by budgeting for large-scale disasters. *In A New Era of Responsibility*, the first FY2010 document issued by the Obama Administration, OMB stated that "in the past, budgets assumed that there would not be any natural disasters in our nation that would necessitate federal help.... This omission is irresponsible, and has permitted past Administrations to project deficits that were lower than likely occur."[29]

Accordingly, the debate over the amount of funding in the DRF generally falls into two basic categories: (1) proponents who advocate the use of emergency supplemental appropriations to augment the DRF, and (2) those who oppose their use.

Proponents of the current system of using emergency supplemental appropriations argue there are several benefits associated with their use. Some of these perceived benefits are summarized as follows:

- Disasters cannot be anticipated, a condition that creates a budgeting challenge. Because of their flexibility, emergency supplemental appropriations facilitate funding decisions for disasters, the timing and severity of which are unpredictable. Budgeting for large disasters through regular appropriations would likely require Congress to reduce funding for other programs to pay for an unknown and possibly non-existent future event.
- The President is authorized to unilaterally determine when federal assistance ismade available after an incident. Congress retains authority to control spendingby voting on supplemental funding legislation. In essence, a balance of powers is at least theoretically maintained through this process.
- A large balance in the DRF may be subject to transfer or rescission.[30] Such actions could carry an additional negative consequence if a large

disaster were to take place after funds have been withdrawn. An emergency supplemental can besized according to the needs of the actual incident.

Arguments used by some of the opponents against the enactment of emergency supplemental appropriations to augment the DRF include the following:

- Supplemental spending allows lawmakers to circumvent budgetary enforcement mechanisms by deliberately underfunding programs in annual appropriations. Because there is a perception that a supplemental appropriation must be passed quickly, supplemental appropriations may be less likely to be scrutinized in the same manner as an annual appropriation.[31]
- Reliance upon emergency supplemental appropriations may result in unnecessarily high funding levels, as early damage estimates may overstate actual needs. Maintaining DRF funds at a level that more accurately reflects the needs would foster fiscal responsibility because all disaster response would have to be budgeted with existing funds.
- Use of emergency supplemental appropriations may give the appearance of fiscal irresponsibility on the part of the federal government, especially when they are frequent and large. Peter Orszag, former Director of the Office of Management and Budget (OMB), voiced this concern when he stated: “The first step in addressing this very deep fiscal hole is honesty. This budget will not play the games that are typically played in which you assume that there will never again be a hurricane or disaster.”[32]
- Emergency supplemental appropriations provide a vehicle for non-germane provisions that may not pass on their own, or make the appropriation contentious, thus slowing down federal recovery assistance after a disaster.

Comparative data on Administration requests, annual appropriations, and supplemental appropriations for the DRF are provided in **Table 1** of this report.

Table 1. Disaster Relief Fund Requests, Appropriations, and Supplemental Appropriations (in millions of dollars)

Fiscal Year	Administrative Request (Nominal)	Administrative Request (2012 Constant Dollars)	Enacted Appropriation (Nominal)	Enacted Appropriation (2012 Constant Dollars)	Supplemental (Nominal)	Supplemental (2012 Constant Dollars)	Total Nominal	Total (Constant 2012 Dollars)
1989	200	333	100	166	1,108	1,845	1,208	2,010
1990	270	434	98	157	1,150	1,847	1,248	2,003
1991	270	418	0	0	0	0	0	0
1992	184	277	185	279	3,993	6,021	4,178	6,296
1993	292	431	292	431	1,735	2,560	2,027	2,989
1994	1,154	1,667	292	422	5,117	7,393	5,409	7,810
1995	320	453	318	450	2,275	3,219	2,593	3,666
1996	320	444	222	308	3,171	4,402	3,393	4,707
1997	320	436	1,320	1,799	3,300	4,498	4,620	6,293
1998	2,708	3,645	320	431	1,600	2,153	1,920	2,582
1999	2,566	3,408	308	409	1,806	2,399	2,114	2,806
2000	2,780	3,621	2,780	3,621	0	0	2,780	3,618
2001	2,909	3,702	1,594	2,028	0	0	1,594	2,027
2002	1,369	1,714	664	831	0	0	664	831
2003	1,843	2,261	800	981	1,426	1,749	2,226	2,729
2004	1,956	2,340	1,800	2,154	2,213	2,648	4,013	4,798
2005	2,151	2,493	2,042	2,366	43,091	49,934	45,133	52,264
2006	2,140	2,398	1,770	1,984	6,000	6,724	7,770	8,702
2007	1,941	2,113	1,500	1,633	4,092	4,454	5,592	6,083
2008	1,652	1,757	1,400	1,489	10,960	11,659	12,360	13,140
2009	1,900	1,996	1,278	1,342	0	0	1,299	1,363
2010	2,000	2,079	1,600	1,664	5,100	5,303	6,700	6,962
2011	1,950	1,986	2,650	2,699	0	0	2,650	2,697
2012	1,800	1,800	700	700	6,400	6,400	7,100	7,095
2013	6,089	6,089	7,007	7,007	11,485	11,485	17,574	17,574
Total	41,084	48,295	31,040	35,351	116,022	136,693	146,165	171,045

Source: CRS analysis of Administration budget documents and appropriations statutes.

Note: FY2012 current dollar conversions were calculated using GDP Chained Price Index data in Table 10.1, FY2014 Budget Historical Tables volume available at http://www.whitehouse.gov/omb/budget/Historicals.

EMERGENCY SUPPLEMENTAL APPROPRIATIONS: FY1989-FY2013

This section provides additional summary information on emergency supplemental appropriations legislation enacted since 1989; it uses a broad concept of what constitutes emergency disaster assistance. The funds cited include appropriations to all federal agencies that undertook disaster relief, repair of federal facilities, and hazard mitigation activities directed at reducing the impact of future disasters. Funds used for activities such as research or administrative costs have been omitted from this analysis in an attempt to focus solely on disaster relief and assistance. Moreover, counterterrorism, law enforcement, and national security appropriations are not included in this compilation. Unless otherwise noted, this report does not take into account rescissions approved by Congress after funds have been appropriated for disaster assistance.

Since FY1989, Congress has appropriated roughly $309 billion for disaster assistance in 37 appropriations acts, primarily supplemental appropriations acts after significant catastrophes occurred in the United States. The mean annual appropriation is approximately $12.9 billion. The mean supplemental appropriation for all 37 bills is approximately $8.3 billion. Current dollar figures for each appropriation and totals for the period appear in **Table 2**.

As reflected in **Table 2**, supplemental appropriations have generally been enacted as stand-alone legislation. In some instances, however, emergency disaster relief funding has been enacted as part of regular appropriations measures, continuing appropriations acts (continuing resolutions), or as a part of omnibus appropriations legislation. Requested funding levels noted in the third column of **Table 2** reflect House Appropriations Committee data on total requested funding for the entire enacted bill.

Recent Enacted Supplemental Appropriations

In the 113th Congress, H.R. 152, the Disaster Relief Appropriations Act, 2013, was introduced in the House by Representative Harold Rogers on January 4, 2013. The bill provided disaster assistance through numerous federal agencies and entities in response to Hurricane Sandy. The bill included provided $11.5 billion for the DRF, $6 million for the Commodity Assistance

Program account—specifically for The Emergency Food Assistance Program (TEFAP), and $13.07 billion for the Department of Transportation. The bill was enacted as P.L. 113-2.[33]

Table 2. Emergency Appropriations for Disaster Relief: All Agencies FY1989 – Present (Dollars in Thousands)

Fiscal Year	Disaster Event and Date of Major Disaster Declaration	Date Signed into Law and P.L. Number	Emergency Assistance Funding	Emergency Assistance Funding (2012 dollars)
2013	Hurricane Sandy, Nov. - Dec. 2012	Jan. 29, 2013, P.L. 113-2	$50,600,000	$50,600,000
2012	Storms, Flooding, Drought, and Hurricane Irene events in 2011	Dec. 23, 2011, P.L. 112-77	$8,124,000	$8,124,000
2010	Hurricane Katrina, severe storms/ flooding, wildfires, oil spill, various dates	July 19, 2010 P.L. 111-212	$5,563,600	$5,784,746
2008	Hurricane Katrina, Midwest Flooding and the 2008 hurricanes, various dates	Sept. 30, 2008 P.L. 110-329	$23,389,800	$24,882,126
2008	Hurricane Katrina and other hurricanes in the 2005 season	June 30, 2008 P.L. 110-252	$8,381,805	$8,916,585
2008	Hurricane Katrina & California Wildfires, Oct. 24, 2007	Nov. 13, 2007 P.L. 110-116	$6,355,000	$6,739,188
2007	Hurricane Katrina, Aug. 29, 2005	May 25, 2007 P.L. 110-28	$7,679,000	$8,358,468
2006	Hurricanes Katrina, Rita, Wilma, Aug. - Sept. 2005 June 15, 2006 P.L. 109-234	$19,340,000	$21,674,267	

Fiscal Year	Disaster Event and Date of Major Disaster Declaration	Date Signed into Law and P.L. Number	Emergency Assistance Funding	Emergency Assistance Funding (2012 dollars)
2006	Hurricanes Katrina, Rita, Wilma, Aug. - Sept. 2005 Dec. 30, 2005 P.L. 109-148	$29,046,985	$32,552,849	
2005	Hurricane Katrina Aug. 29, 2005	Sept. 8, 2005 P.L. 109-62	$51,800,000	$60,025,840
2005	Hurricane Katrina Aug. 29, 2005	Sept. 2, 2005 P.L. 109-61	$10,500,000	$12,167,400
2005	Hurricanes Ivan, Jeanne, Sept. 1, 2004	Oct. 13, 2004 P.L. 108-324	$11,103,887	$12,867,184
2004	Hurricanes Charley, Frances, Sept. 1, 2004	Sept. 8, 2004 P.L. 108-303	$2,000,000	$2,392,979
2004	Wildfires, various dates	Aug. 8, 2004 P.L. 108-287	$500,000	$598,245
2004	Hurricane Isabel Sept. 18, 2003	Nov. 6, 2003 P.L. 108-106	$813,000	$972,746
2003	Storms, various 2003 dates	Sept. 30, 2003 P.L. 108-83	$820,700	$1,006,804
2003	Tornadoes, May 6, 2003	Aug. 8, 2003 P.L. 108-69	$983,600	$1,206,644
2002	Terrorist attacks, Sept. 11, 2001	Aug. 2, 2002 P.L. 107-206	$6,167,600	$7,720,660
2001	Terrorist attacks, Sept. 11, 2001	Sept. 18, 2001 P.L. 107-38	$20,000,000	$25,451,351
2001	Nisqually Earthquake	July 24, 2001 P.L. 107-20	$365,700	$465,378
2000	Hurricane Floyd Sept. 16, 1999	Oct. 20, 1999 P.L. 106-74	$2,480,425	$3,230,658
1999	Tornadoes, various dates	May 21, 1999 P.L. 106-31	$1,296,723	$1,722,424
1999	Hurricanes Georg-es, Bonnie, flood-ing, various dates	Oct. 21, 1998 P.L. 105-277	$1,830,977	$2,432,068

Table 2. (Continued)

Fiscal Year	Disaster Event and Date of Major Disaster Declaration	Date Signed into Law and P.L. Number	Emergency Assistance Funding	Emergency Assistance Funding (2012 dollars)
1998	El Niño floods, Feb. 9, 1998	May 1, 1998 P.L. 105-174	$2,602,173	$3,502,204
1997	Dakotas flooding, Apr. 7, 1997	June 12, 1997 P.L. 105-18	$5,863,883	$7,992,317
1995	Oklahoma City bombing, Apr. 25, 1995 July 27, 1995 P.L. 104-19	$6,599,531	$9,337,651	
1995	Northridge Earthquake, Tropical Storm Alberto, various dates	Sept. 28, 1994 P.L. 103-327	$417,500	$590,719
1994	Midwest floods, CA fires, and Northridge earthquake Jan. 17, 1994	Feb. 12, 1994 P.L. 103-211	$8,837,952	$12,769,849
1993	Midwest floods, June 11, 1993	Aug. 12, 1993 P.L. 103-75	$3,494,750	$5,156,247
1993	Hurricanes Andrew, Iniki, various dates	July 2, 1993 P.L. 103-50	$52,345	$78,930
1992	Hurricanes Andrew, Iniki, Aug. 24, 1992	Sept. 23, 1992 P.L. 102-368	$5,767,116	$8,696,075
1992	L.A. riots/Chicago flood, various dates	June 22, 1992 P.L. 102-302	$469,650	$708,172
1992	Hurricane Bob, various dates	Dec. 12, 1991 P.L. 102-229	$943,000	$1,421,924
1990	Hurricane Hugo, Exxon Valdez, various dates	May 25, 1990 P.L. 101-302	$670,412	$1,076,598

Fiscal Year	Disaster Event and Date of Major Disaster Declaration	Date Signed into Law and P.L. Number	Emergency Assistance Funding	Emergency Assistance Funding (2012 dollars)
1990	Hurricane Hugo, Loma Prieta Earthquake, Oct. 18, 1989	Oct. 26, 1989 P.L. 101-130	$2,850,000	$4,576,746
1989	Hurricane Hugo, Sept. 20, 1989	Sept. 29, 1989 P.L. 101-100	$1,108,000	$1,779,310
1989	Fires on federal lands, various dates	June 30, 1989 P.L. 101-45	$348,969	$581,013
Total			$309,168,083	$358,160,365

Source: Supplemental funding totals derived, in part, from CRS analysis of emergency appropriations after disasters, FY1989-FY2013.

a. Notes: Declaration dates in this table represent the date the President issued a major disaster declaration for the disaster that appeared to be the primary catalyst for the supplemental appropriations legislation. In a series of disasters (such as the Midwest floods of 1993) this date represents the first of several declarations associated with that particular disaster. In some instances, identifying which disasters were primarily associated with consideration of the supplemental appropriations was not possible. FY2012 current dollar conversions were calculated using GDP Chained Price Index data in Table 10.1, FY2014 Budget Historical Tables volume available at http://www.whitehouse.gov/omb/budget/Historicals.

In the 112th Congress H.R. 3672, the Disaster Relief Appropriations Act, 2012, was introduced in the House by Representative Harold Rogers on December 14, 2001. The bill included $6.4 billion for the DRF and roughly $1.7 billion for the Army Corps of Engineers to repair damages to federal projects resulting from major disasters, operations and expenses, and other projects to prepare for floods, hurricanes, and other natural disasters. The bill was enacted as P.L. 112-77 on December 23, 2011.

In the 111th Congress, H.R. 4899, the Disaster Relief and Summer Jobs Act of 2010, was introduced in the House by Representative David Obey on March 21, 2010. The bill included $5.1 billion for the DRF. During consideration H.R. 4899 became a vehicle for additional non-disaster funding including $33 billion in war funding for the Department of Defense, and funding for court case relief for veterans, Native Americans, and minority farmers. The bill was signed into law on July 29, 2010, and became P.L. 111-212.

In addition to the $5.1 billion provided for the DRF, other types of disaster assistance in P.L. 111-212 included $18 million for forest restoration, $49 million for the Economic Development Administration (EDA) for long-term recovery and infrastructure restoration for floods, and another $5 million for technical assistance for activities associated with the Deepwater Horizon oil spill. The act provided $100 million to the Department of Housing and Urban Development (HUD) for housing and economic revitalization, and roughly $26 million was provided to the Department of the Interior to address environmental and fishery impacts resulting from the Deepwater Horizon oil spill.

In the 110th Congress, President George W. Bush signed into law four measures (P.L. 110-28, P.L. 110-116, P.L. 110-252, and P.L. 110-329) that provided roughly $45 billion in supplemental appropriations for disaster relief and recovery (most of it for the DRF). P.L. 110-28, signed on May 25, 2007, included an appropriation of $7.6 billion for disaster assistance, $6.9 billion of which was classified for Hurricane Katrina recovery. P.L. 110-116, signed into law on November 13, 2007, provided a total of $6.3 billion for continued recovery efforts related to Hurricanes Katrina, Rita, and Wilma, and for other declared major disasters or emergencies. P.L. 110-116 also included $500 million for firefighting expenses related to the 2007 California wildfires. P.L. 110-252, signed into law June 30, 2008, provided $8.4 billion in disaster assistance, most of which was directed at continuing recovery needs resulting from the 2005 hurricane season.

P.L. 110-329, signed into law on September 30, 2008, included an appropriation for emergency and disaster relief of $23.4 billion. Of this amount, roughly $2.9 billion was continued disaster relief for the 2005 hurricane season. The largest share of the funding (just over $8.8 billion), however, was for a string of disasters that occurred in 2008 including Hurricanes Gustav and Ike, wildfires in California, and the Midwest floods. One of the largest funding components in P.L. 110-329 was designated for the Department of Housing and Urban Development's (HUD's) Community Development Fund, which received $6.5 billion specifically for disaster relief, long-term recovery, and economic revitalization for areas affected by the 2008 disasters. Other funding in the law included $135 million for wildfire suppression, and a $100 million direct appropriation for the American Red Cross for reimbursement of disaster relief and recovery expenditures associated with emergencies and disasters that took place in 2008.[34]

Supplemental Appropriations for Hurricanes Katrina, Rita, and Wilma

Since FY2005, a number of supplemental appropriations have been enacted to provide the Gulf Coast disaster relief and recovery assistance after Hurricanes Katrina, Rita, and Wilma. The following section provides examples of federal assistance provided in each supplemental appropriation.[35]

Measures Enacted in FY2010

P.L. 111-212 provided FEMA funding for ongoing recovery projects in the Gulf Coast and to pay claims awarded by arbitrators to state, local, and nonprofits for damages resulting from Hurricane Katrina. The claims were the result of damage disputes. In the case of Louisiana's Charity Hospital, FEMA questioned whether all the damages to the aging hospital were storm-related and offered the state $150 million for repairs. The state disagreed and appealed the decision. The arbiters agreed with the state and ruled that FEMA must pay the state $475 million for repairs to the hospital.

Although some of the disaster assistance in P.L. 111-212 funds Gulf Coast recovery, the amount of assistance for projects related to the 2005 hurricane season is unclear. Thus, funding data from P.L. 111-212 have been excluded from the following summary section and **Table 3**.

Measures Enacted in FY2008

On June 30, 2008, Congress enacted the Supplemental Appropriations Act, 2008 (P.L. 110-252). Some of the funding in P.L. 110-252 includes $100 million for the Economic Development Administration's economic development assistance programs, $73 million for the Louisiana Road Home Program, and $300 million for HUD's Community Development fund. The majority of disaster assistance funding (over $6 billion) in P.L. 110-252 was directed to the Corps of Engineers for projects aimed at repairing damages incurred during the 2005 hurricane season, as well as programs designed to mitigate against future hurricanes.[36]

Another supplemental, the Consolidated Security, Disaster Assistance, and Continuing Appropriations Act of 2009, was passed three months later on September 30, 2008 (P.L. 110-329). P.L. 110-329 includes ongoing disaster relief for destruction resulting from the 2005 hurricane season, including $85 million for the Tenant-Based Assistance Program and $15 million for the Public Housing Capital Fund (administered by HUD), and $15 million for school education programs to help local educational agencies with increased homeless students enrollments as a result of the 2005 hurricanes (administered by the Department of Education). The amount provided in the statute for disaster relief as a result of the 2005 hurricane season is roughly $2.9 billion.[37]

Measures Enacted in FY2007

On May 25, 2007, the President signed into law P.L. 110-28, which appropriated $120 billion in emergency supplemental funding for Iraq, Afghanistan, and other matters, including $6.9 billion for continued Gulf Coast relief and waived cost-shares for the Gulf Coast states. The measure was a successor to previous emergency supplemental legislation in the 110th Congress, H.R. 1591, which had been vetoed by the President on May 1, 2007. This was the fifth supplemental measure enacted in the 110th Congress containing disaster assistance specifically provided in response to Hurricanes Katrina and Rita. The sixth supplemental measure enacted as part of P.L. 110-116 on November 13, 2007, provided an additional $6.3 billion for emergency assistance, most, but not all, of which can be attributed to the Gulf Coast recovery. The $3 billion appropriated for HUD's Community Planning and Development Fund can only be used for the Louisiana Road Home program,[38] but the $2.9 billion appropriated for the DRF can be used not only for the Gulf Coast but for other declared disasters as well.

After the enactment of P.L. 110-252, the total amount appropriated by Congress in supplemental funding after the 2005 hurricanes surpassed the $130 billion mark. In addition to these rescissions and appropriations, Congress enacted other funding changes by transferring $712 million from FEMA to the Small Business Administration for disaster loans (P.L. 109-174).

Measures Enacted in FY2006

Later, when Hurricanes Rita and Wilma struck, Congress enacted two additional emergency supplemental appropriations; the costs of both were offset by rescissions. The FY2006 appropriations legislation for the Department of Defense (P.L. 109-148) rescinded roughly $34 billion in funds previously appropriated (almost 70% of which was taken from funds previously appropriated to DHS) and appropriated $29 billion to other accounts primarily to pay for the restoration of federal facilities damaged by the hurricanes).[39] Also in FY2006, Congress agreed to an Administration request for further funding—$19.3 billion was appropriated in supplemental legislation (P.L. 109-234) for recovery assistance, with roughly $64 million rescinded from two accounts ($15 million from flood control, Corps of

Engineers, and $49.5 million from Navy Reserve construction, Department of Defense).

Measures Enacted in FY2005

In response to the widespread destruction caused by three catastrophic hurricanes at the end of the summer of 2005, Congress enacted four emergency supplemental appropriations bills. Two of the bills were enacted as FY2005 emergency supplemental appropriations after Hurricane Katrina devastated parts of Florida and Alabama and resulted in presidential major disaster declarations for all jurisdictions in Louisiana and Mississippi. The two emergency supplemental appropriations (P.L. 109-61 and P.L. 109-62) together provided $62.6 billion for emergency response and recovery needs; most of the appropriations in these two bills funded the DRF.

Emergency Supplemental Appropriations for Hurricanes Katrina, Rita, and Wilma by Federal Agency

The primary recipient of emergency supplemental appropriations related to Hurricanes Katrina, Rita, and Wilma is DHS, which has received 57% of emergency supplemental funding for disaster assistance. HUD received 16%, DOD Army Corps of Engineers received 13%, the DOD received another 7% for miscellaneous activities, and the Department of Transportation received 3%.[40] **Table 3** provides information on the appropriations made in the eight emergency supplemental appropriations enacted largely to address losses associated with Hurricanes Katrina, Rita, and Wilma and identify the departments and agencies that received funding for disaster relief.[41]

ISSUES FOR CONGRESS

Authorizing the DRF

As mentioned earlier in this report, the DRF has not been authorized explicitly by Congress. Authorization measures are generally used to establish, continue, or modify an agency or program for a fixed or an indefinite period of time.

Table 3. FY2005-FY2008 Supplemental Disaster Appropriations after Hurricanes Katrina, Rita, and Wilma (thousands of current dollars)

Department	FY2005		FY2006		FY2007		FY2008		TOTAL
	P.L. 109-61	P.L. 109-62	P.L. 109-148	P.L. 109-234	P.L. 110-28	P.L. 110-116	P.L. 110-252	P.L. 110-329	Department
Agriculture			$1,183,000	$152,000[a]				$38,000	$1,373,000
Department of Commerce			$55,000	$150,000	$110,000		$100,000	$481,000	$896,000
Defense-Military	$500,000	$1,400,000	$5,754,000	$1,488,000[b]					$9,142,000
Defense-Civil/ Corps of Engineers		$400,000	$2,900,000	$3,686,000[c]	$1,433,000		$6,366,988	$1,621,200	$16,407,188
Education and related agencies			$1,600,000	$285,000	$60,000			$30,000	$1,975,000
Health and Human Services			$640,000	$12,000				$600,000[d]	$1,252,000
Homeland Security	$10,000,000	$50,000,000	$285,000	$6,662,000	$4,110,000	$2,900,000		$20,000	$73,977,000
Housing and Urban Development			$11,890,000	$5,200,000	$7,000	$3,000,000	$373,000	$150,000	$20,620,000
Interior			$70,000	$256,000	$10,000	[e]			$336,000
Justice			$229,000	$9,000	$50,000				$288,000
Labor			$125,000	$16,000					$141,000
Transportation			$2,798,000	$702,000[f]	$906,020				$4,406,020
Veterans Affairs			$658,000	$586,000	$14,500				$1,258,500
Armed Forces Retirement Home				$176,000					$176,000
Corp. for National & Community Svc				$10,000					$10,000
Environmental Protection Agency			$8,000	$13,000					$21,000

Department	FY2005		FY2006		FY2007		FY2008		TOTAL
	P.L. 109-61	P.L. 109-62	P.L. 109-148	P.L. 109-234	P.L. 110-28	P.L. 110-116	P.L. 110-252	P.L. 110-329	Department
General Services Administration			$38,000	$37,000					$75,000
Historically Black Colleges Cap. Financing				$15,000					$15,000
Nat'l Aeronautics & Space Admin.			$350,000	$35,000	$20,000				$405,000
Judiciary			$18,000						$18,000
Small Business Administration			$446,000	$542,000	$181,070				$1,169,070
Total	**$10,500,000**	**$51,800,000**	**$29,047,000**	**$20,032,000**	**$6,901,590**	**$5,900,000**	**$6,839,989**	**$2,910,200**	**$133,960,778**

Source: Congressional Research Service. Numbers have been rounded.

a. Does not include authority for $500 million in direct assistance to be drawn from the Commodity Credit Corporation, authorized in Title III of P.L. 109-234.

b. Includes rescissions and military construction accounts.

c. Includes rescissions.

d. This funding amount is divided between the 2005 hurricane season, and hurricanes, floods, and other disasters that occurred in 2008.

e. In Division B of P.L. 110-116, 121 Stat. 1342-1343, Section 157 provides $329 million for Forest Service Wildland Fire Management and $171 million for Bureau of Land Management Wildland Fire Management. This funding is not included in the table since the funding was for wildland firefighting activities and not related to Gulf Coast hurricane relief and recovery.

f. Department of Transportation funds derived from Highway Trust Fund rescission.

The measures are also used to explicitly name accounts or programs.[42] Authorizations also establish the duties and functions of an agency or program, its organizational structure, and the responsibilities of agency or program officials. One function of such legislation is to authorize subsequent spending in appropriations bills for specific agencies, programs, projects, accounts, or funds; such authorizations may include spending ceilings.[43]

It may be argued that a large account such as the DRF should be authorized specifically by name to ensure these functions are formally established. Others may view this as unnecessary because they believe the Stafford Act, which authorizes appropriations to carry out disaster relief, preparedness, and hazard mitigation activities, is sufficient.[44]

If authorizing the DRF became an issue of active debate, Congress could consider legislation similar to Section 329 of H.R. 3377 (introduced in the 111th Congress) to authorize the DRF, or it could elect to maintain the DRF as described under 42 U.S.C. 5197e of the Stafford Act.

Restructuring Budgetary Procedures

Some proposals have been advanced to reduce the need for emergency supplemental appropriations. For example, the Bipartisan Task Force on Funding Disaster Relief[45] offered several policy options for changing budgetary procedures for funding disaster assistance. Some of these options for restructuring the budgetary process include the following:

- **Eliminate any Nonemergency Funding in a Bill.** Many emergency supplemental bills include a variety of funding and other needs in a single bill. Eliminating nonemergency (or nongermane) elements from legislation could prevent the passage of legislation that might not pass Congress if it were not attached to an emergency supplemental appropriation. Such action may make emergency supplemental appropriations less controversial and ensure expedited enactment.
- **Strengthening Criteria.** The Bipartisan Task Force offered the use of more strict criteria as a policy option. If specific criteria were enforced, it is argued, Congress and the President would have to issue written justifications for designating appropriations as emergencies. Proponents of rewriting the criteria say the measure would open the debate on the use of emergency supplemental appropriations by

adding an additional layer of scrutiny to the emergency appropriation process. Opponents may argue that changing the criteria may create unnecessary delays for appropriating needed emergency funding.

- **Increase Funding to the DRF.** As mentioned earlier in this report, Congress may decide to increase the funding level of the DRF through annual appropriations. Doing so could eliminate, or at least greatly decrease, the need for emergency supplemental funding.
- **The Creation of a Rainy-Day Fund.** A rainy-day fund, also known as a reserve account, would be financed by cuts in other discretionary accounts, or through revenue raising measures. Spending from this fund would then be allowed only when needed for expensive disasters. Proponents of this policy option would likely argue a rainy-day fund carries several advantages. For one, in contrast to emergency supplemental appropriations which increase the federal deficit through borrowing funds, rainy-day accounts do not add to the federal deficit because they are funded through savings and/or revenue raising measures. Furthermore, the balance for a rainy-day fund would increase during periods in which there are few, or relatively small disasters. Another advantage would be that a rainy-day fund would not require a restructuring of the process of administering disaster relief.
- **The Creation of a Contingency Fund.** A contingency fund based on a cost analysis of previous disasters could be created for use after large a disaster occurs. A contingency fund could be funded at a level sufficient for large disasters, while relatively routine disasters would still be funded through the DRF. Unlike a rainy-day fund—which pays for disasters through savings and revenue generating measures—a contingency fund would receive an annual appropriation. As the contingency fund is drawn down, the account would be replenished through regular appropriations, obviating the need for emergency supplemental appropriations. During congressional testimony, James Lee Witt, who was the Director of FEMA during the Clinton Administration, summarized the fund as follows:

> What we did and what worked so well with Congress and with OMB was we were able to use that five year cost analysis to set up a contingency fund that was under OMB that was budgeted, that you did not have to go back for supplementals every time you had a disaster declaration. And then all we would do then is go to OMB and say we

have a Presidential disaster declaration that's going to cost $400 million.[46]

- **Model Federal Disaster Funding on State Statutes.** In the majority of states, the first recourse for funding disaster relief is to reallocate departmental appropriations to meet emergency needs. Should the reallocation prove insufficient, the legislature may be called into special session to approve the appropriation of additional funds. In some states, local agency funds must be exhausted before state financial assistance is requested.

In other states, state statutes impose limits on the amount that may be expended on a given disaster. For example, the governor of Alaska may use up to $1 million in state funds per year for disaster assistance. Another example is Hawaii, which caps the amount of funding for a single disaster at $1 million.[47]

Proponents of applying the model used by some states may claim that such measures would reduce the amount needed to fund the DRF. Opponents of the measure, on the other hand, may claim that reallocating funds from other accounts would jeopardize programs that already struggle with tight budgets. They may also argue that should the reallocation of funds for disaster relief become a standard practice, there would be a need to fund potential donor accounts at higher levels, defeating the purpose of reallocating funds to save money. Finally, they may argue that its usage would be impractical for large events such as Hurricane Katrina, which would still require an emergency supplemental appropriation for disaster relief.

If there is interest in changing current practices in funding disaster relief, Members of Congress may wish to further explore other policies for reforming the disaster funding budgeting process such as the ones discussed above, or formulate other strategies that may potentially reform federal disaster relief funding.

Additional Approaches for Funding Federal Disaster Relief

Should Congress become concerned that disaster relief negatively affects the federal deficit, or that the federal share of disaster relief has become disproportionate, other avenues exist for decreasing the amount of federal funding. Two of these include the following:

- **Decrease the Federal Cost Share.** Generally, when the President declares a disaster the federal government reimburses 75% of a state's disaster relief expenses. The cost share could be altered so that states pay for more of their disaster activities. For example, prior to the Midwest floods in 1993, the federal cost share for mitigation was 50%. Reducing the federal cost share could significantly decrease the amount the federal government pays for disaster relief.
- **Strengthen Declaration Criteria.** It may be argued that the current declaration process allows what some call "marginal" disasters to receive federal funding. Amarginal disaster is an event that some argue could be handled by the state without federal assistance. A set of standardized criteria could be developed to ensure that only events meeting a certain criteria would be eligible for federal disaster relief. Proponents would further argue that scaling back federal fundingmay be beneficial because states may trim investments in mitigation if federal assistance is too generous.[48]

Oversight on Reporting

A 2006 Government Accountability Office (GAO) report indicated there was a need to improve the information in FEMA's weekly reports on the status of hurricane relief, and that OMB should take action to improve transparency and accountability regarding the status of hurricane related funding.[49] Both OMB and FEMA agreed that these improvements were needed and would be forthcoming. Congress could authorize oversight mechanisms to investigate the extent to which FEMA has made such improvements. For example, Section 203 of H.R. 5351 (introduced in the 109th Congress) would have required each state, local, tribal, and non-profit entity that received federal assistance funds in response to catastrophic events or other emergencies to report to the pertinent federal agency six months after the initial disbursement of resources. Furthermore, the legislation would have also required any agency that disbursed federal assistance funds to report to the Inspector General of the Department the purpose for which resources were provided, the amounts disbursed, allocated, and expended, and the status of reporting by agencies that received disbursements.

CONCLUDING POLICY QUESTIONS

Since the 1950s, the level of financial assistance given to states for disaster relief by the federal government has steadily increased. In light of the federal deficit, the increased federal involvement has raised policymaking questions concerning how disaster relief should be equitably funded. Some of these questions include the following:

- The model for emergency and disaster response is built on the premise that emergencies and disasters are local. Requests for assistance from the next level of government are made only if that unit of government is overwhelmed. Some would argue that some incidents funded by the federal government do not meet this requirement. An example might be snow removal or repairs after minor flooding. Is the federal government funding emergencies and major disasters that do not meet the criterion of the states being overwhelmed before requesting assistance? Are states using federal funding for disaster relief to protect their budgets?
- Should federal disaster relief be subject to thresholds and maximums? For example, an emergency or major disaster might not receive federal funding unless damage estimates reach a certain level. As another example, the total amount of federal relief for an event could be capped at a certain amount. After this level has been reached, the state would then be responsible to pay for the rest of recovery.
- Should the state's fiscal capability factor into disaster relief? In 1986 FEMA proposed measures to reduce the amount the agency contributed toward disaster relief. One of the proposals argued that funding allocations should be made according to each state's ability to fund its own disaster relief. The determination would be based on a comparison of the state's per capita income with the national per capita income.[50] The calculation would then be used to create a sliding scale for assistance. Communities capable of funding their own disaster relief would receive limited or no assistance. In contrast, struggling communities would be eligible to receive more federal assistance.[51]
- Some may argue that federal funding for disaster relief has become entrenched to the point that it has contributed to unintended consequences. For example, it has been argued that some states do not properly fund mitigation measures because there is a presumption that

federal funding is virtually guaranteed should an emergency or major disaster occur.[52] Those advocating this position could arguably point out that federal involvement in disaster relief will continue to increase and that in order to be fiscally responsible, changes should be made in the way in which disaster relief is funded. Others may claim the function of the federal government is to help states in their time of crisis. Withholding, or limiting the amount of funding a state could receive for an incident would be neglectful of that state's needs.

- OMB's *A New Era of Responsibility* projects that spending for disaster costs for FY2010 will be $11 billion. By 2019 disaster costs are projected to rise to $30 billion.[53] This represents an increase of 173% in disaster costs. On what basis did the Administration calculate this increase in disaster costs?
- Congress requires FEMA to submit a monthly status report on the DRF.[54] The reports must detail obligations, allocations, and expenditures for Hurricanes Katrina, Rita, and Wilma. Other than the DRF report, scant data exist on other federal funding for emergencies and major disasters. In light of the amount of federal funding going to these incidents, could better transparency be achieved by requiring the same level of reporting for all declared emergencies and major disasters?

These and other questions may be raised should Congress elect to debate the past and future funding of disaster relief.

End Notes

[1] 42 U.S.C. 5121 et seq. For information on the declaration process, see CRS Report RL34146, *FEMA's Disaster Declaration Process: A Primer*, by Francis X. McCarthy.

[2] Per 42 U.S.C. 5122(4), the District of Columbia, the U.S. territories, and tribal governments use the same declaration process.

[3] U.S. Federal Emergency Management Agency. "The Disaster Process and Disaster Aid Programs." http://www.fema.gov/hazard/dproc.shtm.

[4] A structure does not have to be damaged to be eligible for mitigation. Hazard Mitigation Grants are available as block grant funds and can be used outside of the disaster area. For more information, see CRS Report R40471, *FEMA's Hazard Mitigation Grant Program: Overview and Issues*, by Natalie Keegan.

[5] 44 C.F.R. § 206.101(2).

[6] U.S. Congress, Senate Task Force on Funding Disaster Relief, *Bipartisan Task Force on Funding Disaster Relief*, 104th Cong., 1st sess., March 15, 1995, No. 104-4 (Washington: GPO, 1995), p. 1.

[7] 62 Stat. 1031.

[8] See 42 U.S.C. 5202 (note).

[9] P.L. 100-707, 102 Stat. 4708.

[10] Keith Bea, "The Formative Years: 1950-1978," in *Emergency Management: The American Experience*, ed. Claire B. Rubin (Fairfax, VA: Public Entity Risk Institute, 2007), p. 81.

[11] Richard Sylves, *Disaster Policy and Politics: Emergency Management and Homeland Security* (Washington: CQ Press, 2008), p. 48.

[12] P.L. 81-920.

[13] P.L. 81-875, 64 Stat. 1109.

[14] P.L. 89-796, 80 Stat. 1316.

[15] P.L. 93-288, 88 Stat. 143.

[16] 102 Stat. 4689.

[17] Richard Sylves, *Disaster Policy and Politics: Emergency Management and Homeland Security* (Washington: CQ Press, 2008), p. 49.

[18] Richard Sylves, *Disaster Policy and Politics: Emergency Management and Homeland Security* (Washington DC: CQ Press, 2008), p. 54.

[19] Section 1110 the Sandy Recovery Improvement Act of 2013 (P.L. 113-2) amended Sections 401 and 501 of the Stafford Act to allow tribal governments to request disaster assistance from the federal government. Prior to the act the state had to request assistance. For more information on the Sandy Recovery Improvement Act of 2013 see CRS Report R42991, *Analysis of the Sandy Recovery Improvement Act of 2013*, by Jared T. Brown, Francis X. McCarthy, and Edward C. Liu.

[20] CRS Report RL33053, *Federal Stafford Act Disaster Assistance: Presidential Declarations, Eligible Activities, and Funding*, by Francis X. McCarthy.

[21] Title VI of P.L. 109-295, the FY2007 DHS appropriations legislation.

[22] §528, P.L. 109-295, 120 Stat 1383.

[23] A similar reporting requirement has been carried forward in subsequent appropriations statutes. See, for example, P.L. 110-329, 122 Stat 3674-75.

[24] This section is based on the author's in-person interview with a Department of Homeland Security (DHS) finance official, March 9, 2009.

[25] The DRF is a "no year" account; funds remain until expended.

[26] An example of a duplication of a benefit fund is when the state receives an insurance payment for disaster damages after obtaining federal funding for the same damages. In such a case the state is to return the funding to the federal government.

[27] Open incidents are those whose related activities are currently being funded by FEMA.

[28] Based on U.S. Government Accountability Office, *Disaster Relief: Government Framework Needed to Collect and Consolidate Information to Report on Billions in Federal Funding for the 2005 Gulf Coast Hurricanes,* GAO-06-834, September 2006. The Department of Defense (DOD) is the largest recipient of supplemental funding.

[29] Office of Management and Budget, *A New Era of Responsibility: Renewing America's Promise*, Washington DC, February 26, 2009, p. 36.

[30] For more information on rescissions and disaster relief see CRS Report R42458, *Offsets, Supplemental Appropriations, and the Disaster Relief Fund: FY1990-FY2013*, by William L. Painter.

[31] U.S. Congress, Congressional Budget Office, *Supplemental Appropriations in the 1990s*, March 2001, http://www.cbo.gov/ftpdocs/27xx/doc2768/EntireReport.pdf.

[32] CQ Financial Transcripts, "OMB Director Orszag and CEA Chair Romer Hold Briefing on the Fiscal 2010 Budget Request," press release, February 26, 2009, http://www.cq.com/display.do?dockey=/cqonline/prod/data/docs/html/transcripts/financial/111/financialtranscripts111-000003061915.html@committees&pub=financialtranscripts&print=true#speakers. The President's request for the DRF for FY2010 was increased by $600 million over the FY2009 enacted amount of $1.4 billion.

[33] For more information on supplemental funding for Hurricane Sandy see CRS Report R42869, *FY2013 Supplemental Funding for Disaster Relief*, coordinated by William L. Painter and

Jared T. Brown. See also CRS Report R42803, *Federal Involvement in Flood Response and Flood Infrastructure Repair: Hurricane Sandy Recovery*, by Nicole T. Carter.

[34] Congress did not meet the full request of the American Red Cross, which requested $150 million for reimbursement of disaster relief and recovery expenditures as a result of disasters occurring in 2008. This is not the first time Congress appropriated funds for the organization. In 2004, Congress gave $70 million in aid to the American Red Cross after four hurricanes hit Florida (118 Stat, 1251-1252).

[35] For information on federal expenditures for Hurricanes Katrina, Rita, Wilma, Gustav, and Ike see CRS Report R43139, *Federal Disaster Assistance after Hurricanes Katrina, Rita, Wilma, Gustav, and Ike*, coordinated by Bruce R. Lindsay and Jared C. Nagel.

[36] See CRS Report RL33188, *Protecting New Orleans: From Hurricane Barriers to Floodwalls*, by Nicole T. Carter.

[37] Some of the funding in P.L. 110-329 was directed at the 2005 hurricanes and hurricanes, flooding, and other disasters occurring in 2008. An exact amount for each of these events could not be identified from the legislative text.

[38] See CRS Report RL34410, *The Louisiana Road Home Program: Federal Aid for State Disaster Housing Assistance Programs*, by Natalie Keegan.

[39] In requests to Congress, President Bush termed the sequence of events as a "reallocation" of funds.

[40] See **Table 3**.

[41] For more information on disaster assistance expenditures for Hurricanes Katrina, Rita, Wilma, as well as information on Gustav and Ike CRS Report R43139, *Federal Disaster Assistance after Hurricanes Katrina, Rita, Wilma, Gustav, and Ike*, coordinated by Bruce R. Lindsay and Jared C. Nagel.

[42] CRS Report RS20371, *Overview of the Authorization-Appropriations Process*, by Bill Heniff Jr.

[43] A ceiling arguably limits the amount of funding an account may receive. The ceiling however, is not binding, Congress may appropriate funds at levels above the authorization.

[44] 42 U.S.C. 5197e.

[45] U.S. Congress, Senate Bipartisan Task Force on Funding Disaster Relief, *Federal Disaster Assistance*, Report of the Senate Task Force on Funding Disaster Relief, 104th Cong., 1st sess., March 15, 1995, No. 104-4 (Washington: GPO, 1995), pp. 69-74.

[46] U.S. Congress, House Committee on Small Business, *Hearing on Disaster Relief and Access to Capital Legislation*, 110th Cong., 1st sess., March 8, 2007, 110-6.

[47] CRS Report RL32287, *Emergency Management and Homeland Security Statutory Authorities in the States, District of Columbia, and Insular Areas: A Summary*, by Keith Bea, L. Cheryl Runyon, and Kae M. Warnock.

[48] For more information on changing declaration criteria see CRS Report R42702, *Stafford Act Declarations 1953-2011: Trends and Analyses, and Implications for Congress*, by Bruce R. Lindsay and Francis X. McCarthy. See also James F. Miskel, *Disaster Response and Homeland Security: What Works, What Doesn't* (Westport, CT: Praeger Security International, 2006), p. 126.

[49] U.S. Government Accountability Office, *Disaster Relief: Governmentwide Framework Needed to Collect and Consolidate Information to Report on Billions in Federal Funding for the 2005 Gulf Coast Hurricanes*, GAO-06-834, September 2006.

[50] U.S. Congress, House Committee on Public Works and Transportation, Subcommittee on Investigations and Oversight, *The Federal Emergency Management Agency's Proposed Disaster Relief Regulations (Budget Driven Rulemaking)*, committee print, 100th Cong., 1st sess., August 1987, 75-963 (Washington: GPO, 1987), pp. 4-5.

[51] Under current law (42 U.S.C. 5163) areas cannot be precluded from receiving assistance *solely* on the basis of a sliding scale or formula. Congress amended the Stafford Act in this fashion as a response to the 1986 proposed regulation.

[52] See James F. Miskel, *Disaster Response and Homeland Security: What Works, What Doesn't* (Westport, CT: Praeger Security International, 2006), or Rutherford H. Platt, *Disasters and Democracy: The Politics of Extreme Natural Events* (Washington DC: Island Press, 1999) for arguments against increased federal involvement in disaster assistance.

[53] Office of Management and Budget, *A New Era of Responsibility: Renewing America's Promise*, Washington DC, February 26, 2009, p. 117.

[54] P.L. 110-161, also 42 U.S.C. 5208.

In: FEMA's Disaster Relief Fund
Editor: Angelica Conner
ISBN: 978-1-63321-630-3

Chapter 3

FEMA's Disaster Declaration Process: A Primer*

Francis X. McCarthy

Summary

The Robert T. Stafford Disaster Relief and Emergency Assistance Act (referred to as the Stafford Act - 42 U.S.C. 5721 et seq.) authorizes the President to issue "major disaster" or "emergency" declarations before or after catastrophes occur. Emergency declarations trigger aid that protects property, public health, and safety and lessens or averts the threat of an incident becoming a catastrophic event. A major disaster declaration, issued after catastrophes occur, constitutes broader authority for federal agencies to provide supplemental assistance to help state and local governments, families and individuals, and certain nonprofit organizations recover from the incident.

The end result of a presidential disaster declaration is well known, if not entirely understood. Various forms of assistance are provided, including aid to families and individuals for uninsured needs and assistance to state and local governments and certain non-profits in rebuilding or replacing damaged infrastructure.

The amount of assistance provided through presidential disaster declarations has exceeded $140 billion. Often, in recent years, Congress

* This is an edited, reformatted and augmented version of a Congressional Research Service publication, CRS Report for Congress RL34146, from www.crs.gov, dated May 18, 2011.

has enacted supplemental appropriations legislation to cover unanticipated costs. While the amounts spent by the federal government on different programs may be reported, and the progress of the recovery can be observed, much less is known about the process that initiates all of this activity. Yet, it is a process that has resulted in an average of more than one disaster declaration a week over the last decade.

The disaster declaration procedure is foremost a process that preserves the discretion of the governor to request assistance and the President to decide to grant, or not to grant, supplemental help. The process employs some measurable criteria in two broad areas: Individual Assistance that aids families and individuals and Public Assistance that is mainly for repairs to infrastructure. The criteria, however, also considers many other factors, in each category of assistance, that help decision makers assess the impact of an event on communities and states.

Under current law, the decision to issue a declaration rests solely with the President. Congress has no formal role, but has taken actions to adjust the terms of the process. For example, P.L. 109-295 established an advocate to help small states with the declaration process. More recently, Congress introduced legislation, H.R. 3377, that would direct FEMA to update some of its criteria for considering Individual Assistance declarations.

Congress continues to examine the process and has received some recommendations for improvements. Given the importance of the decision, and the size of the overall spending involved, hearings have been held to review the declaration process so as to ensure fairness and equity in the process and its results.

BACKGROUND

Under the Robert T. Stafford Disaster Relief and Emergency Assistance Act (P.L. 93-288) there are two principal forms of presidential action to authorize federal supplemental assistance. Emergency declarations are made to protect property and public health and safety and to lessen or avert the threat of a major disaster or catastrophe.[1] Emergency declarations are often made when a threat is recognized (such as the emergency declarations for Hurricane Katrina which were made prior to landfall) and are intended to supplement and coordinate local and state efforts prior to the event such as evacuations and protection of public assets. In contrast, a major disaster declaration is made as a result of the disaster or catastrophic event and constitutes a broader authority that helps states and local communities, as well as families and individuals, recover from the damage caused by the event.[2] The differences between the two forms of declarations remain an area of study regarding what events may or may not qualify for the respective declarations.[3]

Federal disaster assistance has served as the impetus for many supplemental appropriations bills over the last several decades, and has accounted for presidential declarations in every state and territory.[4] The supplemental funds are placed in the Disaster Relief Fund (DRF) which is a "no-year" fund managed by the Federal Emergency Management Agency (FEMA) and used only for spending related to presidentially declared disasters.

Major disasters can be a dominant story in the mass media that captures attention both for the devastation that results as well as the potential help that is expected. As one observer noted:

> Disaster assistance is an almost perfect political currency. It serves humanitarian purposes that only the cynical academic could question. It is largely funded out of supplemental appropriations and thus does not officially add to the budget deficit. It promotes the local economy of the area where the building process occurs.[5]

While disaster assistance may be good "political currency," a disaster declaration is generally the result of a tragic and devastating incident that disrupts (and sometimes takes) the lives of hundreds or thousands of families and individuals and the communities and states where they reside. The long-term economic and environmental impact of a disaster can be severe. The assistance offered from federal and private sources may or may not be commensurate with the damage inflicted by a natural or man-made event. Following a disaster, years of rebuilding and recovery work may lie ahead for communities and states. It is the declaration process that sets the federal recovery help in motion.

The trigger for federal disaster assistance is contained in a relatively short statutory provision. P.L. 93-288 (the Stafford Act) includes one brief section that establishes the legal requirements for a major disaster declaration:

> Section 401. Procedures for Declaration. All requests for a declaration by the President that a major disaster exists shall be made by the Governor of the affected state. Such a request shall be based on a finding that the disaster is of such severity and magnitude that effective response is beyond the capabilities of the state and the affected local governments and that the federal assistance is necessary. As a part of such request, and as a prerequisite to major disaster assistance under this

Act, the Governor shall take appropriate response action under state law and direct execution of the state's emergency plan. The Governor shall furnish information on the nature and amount of State and local resources which have been or will be committed to alleviating the results of the disaster and shall certify that, for the current disaster, state and local government obligations and expenditures (of which state commitments must be a significant proportion) will comply with all applicable cost-sharing requirements of this Act. Based on the request of a Governor under this section, the President may declare under this Act that a major disaster or emergency exists.[6]

The process for an emergency declaration is also contained in the Stafford Act. In part it is similar to a major disaster declaration in finding and process, but the actual authorities are limited. In addition, section (b) provides for a special authority for the President to exercise his discretion for events that have a distinctly federal character:

Sec. 5191. Procedure for declaration. (a) Request and declaration All requests for a declaration by the President that an emergency exists shall be made by the Governor of the affected State. Such a request shall be based on a finding that the situation is of such severity and magnitude that effective response is beyond the capabilities of the State and the affected local governments and that Federal assistance is necessary. As a part of such request, and as a prerequisite to emergency assistance under this chapter, the Governor shall take appropriate action under State law and direct execution of the State's emergency plan. The Governor shall furnish information describing the State and local efforts and resources which have been or will be used to alleviate the emergency, and will define the type and extent of Federal aid required. Based upon such Governor's request, the President may declare that an emergency exists. (b) Certain emergencies involving Federal primary responsibility. The President may exercise any authority vested in him by section 5192 of this title or section 5193 of this title with respect to an emergency when he determines that an emergency exists for which the primary responsibility for response rests with the United States because the emergency involves a subject area for which, under the Constitution or laws of the United States, the United States and authority. In determining whether or not such an emergency exists, the President shall consult the Governor of any affected State, if practicable. The President's determination may be made without regard to subsection (a) of this section.[7]

The declaration process is elaborated upon in regulations, specifically in Subpart B of Part 206 (44 CFR). While these regulations have been adjusted through the regulatory process during the past three decades, since 1974 the procedures have undergone little significant change. The process itself is representative of the historical progression of federal disaster relief from being of an episodic nature to the current commonplace disaster declaration, now occurring on a weekly basis. The context in which disaster relief has grown has been in keeping with the growth of government and its concerns.

> Federal disaster relief has a long history in the U.S. dating back to the last years of the eighteenth century and arguably provided much of the political genesis for the New Deal social welfare programs (Landis 1999; Landis 1998; Moss 1999). As Michele Landis argues, social and political construction of claimants for relief as helpless victims of external forces beyond their control ("Acts of God") have exerted an enduring influence on American political discourse, which has manifested itself in heavy reliance on prior political precedents and analogies in constructing responses to current disasters.[8]

Information contained in the Preliminary Damage Assessments (PDAs) and summaries prepared by FEMA regional offices that accompany gubernatorial requests were long considered "pre-decisional and deliberative information" by the executive branch because they are part of the package that is developed and sent to the White House for the President's review and ultimate decision. These materials have generally not been available under the Freedom of Information Act process.[9] However, during 2008, FEMA began to list on its website the results of the PDAs. While the summaries of FEMA recommendations are still not available, the agreed upon figures from the PDAs are now available to the public for review.[10]

Policy makers have found it difficult to achieve equity in the treatment of disparate natural disaster events. The events can vary widely in their type, scope, duration, and impact. Perhaps the greatest variables are the states they affect. Each state has a different topography, a different history, and different capacities to respond and recover based on their own authorities, resources, and choices in what they will do following a disaster. Given those variations, it is a daunting task to construct a uniform process that can account for the range of natural and governmental circumstances that are a part of the nation's potential disaster landscape.

Congress and the Declaration Process

The Impetus for Reform

The question of how the federal disaster declaration process should work has come to the attention of Congress from time to time. Congress has requested reports from its investigative arms and investigated the process in panels that have considered aspects of the disaster relief process.[11] Various administrations have also considered the process and FEMA's role in it.

In 1981, and again in 2001, GAO issued reports on the declaration process that questioned the quality and consistency of FEMA's assessment criteria as well as the agency's ability to produce valid recommendations to the President on a governor's request for supplemental aid. As the most recent report (2001) concluded:

> These criteria are not necessarily indicative of a state's ability to pay for the damage because they do not consider the substantial differences in states' financial capacities to respond when disasters occur. As a result, federal funds may be provided for some disasters when they are not needed – a result that would be inconsistent with the Stafford Act's intent.[12]

Congressional interest in the declaration process derived, in part, from the increased cost in emergency spending—a recurring subject for the appropriations committees when considering supplemental spending legislation.[13] One review of disaster funding, particularly supplemental appropriations, found that the great majority of Disaster Relief Fund spending is related to the large, catastrophic events that meet the criteria established in the Stafford Act. These supplementals have been "driven" by the urgency of large natural disasters. As one witness explained in congressional testimony:

> Since funds provided in annual appropriations measures may be used to cover the federal costs of less catastrophic major disasters, Congress has focused critical attention on "must pass" legislation that would replenish the Disaster Relief Fund for the worst and most costly disasters.[14]

A review of data for a seven-year period from 1988 to 1995 reveals that large expenditures, as funded by supplemental bills, relate to declarations issued for the largest events.[15] During this time period, disaster declarations were made for Hurricane Hugo, the Loma Prieta earthquake, Hurricane Andrew, the Midwest floods of 1993, and the Northridge earthquake. However, these were not the only events deemed worthy of presidential action and of cost to the federal treasury. As summarized by one author:

> But like the tail of a comet, over 200 other declarations accounted for one quarter of such outlays, many of them of relatively minute cost and extent. While of lesser impact on the national treasury, such "low end" declarations have become, to some observers, new sources of federal spending at the local level, long referred to in other contexts as "pork barrel spending."[16]

During early 2001, the term "entitlement" was beginning to be used in describing federal disaster spending by the new Bush Administration. Underlining this thinking in his first appearance before the Senate Appropriations Committee, FEMA Director Joe M. Allbaugh explained:

> FEMA is looking at ways to develop a meaningful and objective criteria for disaster declarations that can be applied consistently. These criteria will not preclude the President's discretion but will help states better understand when they can reasonably turn to the federal government for assistance and when it would be more appropriate for the state to handle the disaster itself.[17]

During the 110th Congress, attention was given to the disaster declaration process and what states and local governments can reasonably expect from that process. During the early spring of 2007, there were tornadoes that had an impact on Arkansas, Alabama and Georgia. The latter two states received disaster declarations while the damage in Arkansas was deemed insufficient to warrant federal assistance. Following these actions, Chairman Bennie Thompson of the House Homeland Security Committee scheduled a hearing to review FEMA's procedures. Representative Thompson set the context as follows:

> As Members representing real communities back home, we want to understand just how FEMA makes determinations regarding what is a disaster deserving of attention and when folks have to fend for themselves. Quite simply, we must have a serious discussion on what our expectations are of our federal government and what should remain a state and local responsibility.[18]

In order to examine the process and the financial projections of disaster spending in particular, Congress mandated in P.L. 110-28 that GAO study how FEMA develops its estimates for supplemental funding that may be needed.[19] Report language detailed these instructions for the review:

> The Committee continues to be concerned with FEMA's ability to manage resources in a manner that maximizes its ability to effectively and efficiently deal with disasters. One aspect of particular concern is how FEMA makes projections of funding needed in response to any given disaster or to meet future disasters. A recent Government Accountability Office (GAO) report raised concerns about FEMA's ability to manage its day-to-day resources and the lack of information on how FEMA's resources are aligned with its operations. As a follow-up to this report, the Committee requests that within six months of enactment GAO review how FEMA develops its estimates of the funds needed to respond to any given disaster. Such review should include how FEMA makes initial estimates, how FEMA refines those estimates within the first few months of a disaster, and how closely FEMA's estimates predict actual costs. The review should also include additional analysis and recommendations regarding FEMA's ability to manage disaster-related resources in a manner that maximizes effective execution of its mission.[20]

The study parameters mandated by Congress are broader than the declaration process, placing its emphasis on ongoing FEMA spending, including the refinement and revision of early estimates as well as the general management of spending projections for the Disaster Relief Fund (DRF). But it recognizes the importance of the initial estimates in decision-making and that the declaration process represents the fundamental decision to both establish federal participation following a disaster and provides the initial estimated amount of resources needed that informs that decision.

GAO's eventual report in response to this request suggested that FEMA's estimates for disaster costs have improved but could use additional refinement; for example

> some of the factors that can lead to changes in FEMA's cost estimates are beyond its control, such as the discovery of hidden damage. Others are not, such as its management of mission assignments. Sensitivity analyses to identify the marginal effect of key cost drivers could provide FEMA a way to isolate and mitigate the effect of these factors on its early estimates. To better predict applicant costs for the Individual Assistance program, FEMA could substitute or add more geographically specific indicators for its national average.[21]

During the 111th Congress, a hearing before the House Homeland Security Committee focusing on FEMA's regional offices also elicited comments regarding the relative speed, and perceived lack of transparency, of the declaration process. The comments came from both a committee member and a state official closely involved in the process.

> Ranking Republican Mike D. Rogers of Alabama questioned the timeliness of FEMA's disaster declarations. Declarations that should take a few days often take a month or longer, and a lack of available information on the status of requests only adds to state officials' frustration, said Brock Long, director of Alabama's Emergency Management Agency. "When we make phone calls to the region or to headquarters, a lot of time the answer we get is 'the declaration request is in process' and that's it," Long said. [22]

The Skepticism of Reform

There have been several proposals to impose more stringent regulations on the declaration process in an attempt to more precisely assess disaster impacts, calculate eligible damage, and incorporate some measure of suffering and loss. Differing perceptions of the declaration process resulted in different reactions to these proposed reforms.

The history of the disaster declaration process is rife with reform efforts that were perceived by some not as reform but as punitive measures directed at certain constituencies. Some of those differing perceptions were held by those closest to the event at the local level that had experienced a disaster. A different perception was also often held by governors that wanted to protect their option to request federal help. These perceptions were also shared, in some instances, by Members of Congress representing affected areas. All elected officials argued, in various forms, that reform should not impede the delivery of needed federal aid.

One example, now in law, of the desire to reform but not obstruct the declaration process is in the Post-Katrina Emergency Management Reform Act of October of 2006. That statute created a new position at FEMA: the Small State and Rural Advocate.[23] One of the principal duties of the advocate is to ensure that the needs of smaller states and rural communities will be "met in the declaration process;" the advocate is also directed to "help small states prepare declaration requests, among other duties."[24]

Another example was FEMA's response to recommendations from GAO's 1981 report. FEMA drafted regulations in 1986 that would be more certain in their delineation of state requirements prior to a declaration, reduce overall federal contributions, and install a formula to determine whether a state would receive a presidential disaster declaration for repairs to state and local infrastructure (known as "public assistance" in the Stafford Act). As the congressional report on the initiative summarized the FEMA draft:

> The proposal would have limited the number of future presidential declarations by establishing a "state deductible" based on a per capita minimum dollar amount adjusted by the ratio of the state/local price index to the national index. Of 111 declarations issued in prior years, 61 would have been ineligible for any public assistance. FEMA also proposed to decrease the federal share for disaster costs from 75 percent to 50 percent and to exclude aid to special districts.[25]

The FEMA proposal of 20 years ago addressed some problems identified by students and critics of the declaration process. However, some Members of Congress viewed this proposal as a means of removing the power of discretion from elected leadership. Rather than apply the empirical solution suggested by FEMA to perceived problems in the declaration process, Congress instead legislated a provision to explicitly forbid the primacy of any "arithmetic formula." In place of agreeing to the regulatory changes in the formula proposed by FEMA and the Administration, Congress added the following section to the Stafford Act:

> Limitation on the Use of Sliding Scales. Section 320. No geographic area shall be precluded from receiving assistance under this Act solely by virtue of an arithmetic formula or sliding scale based on income or population.[26]

While enactment of this legislation halted FEMA's efforts, the issue did not disappear. As discussed below, FEMA has since adopted regulations that use, but not "solely", arithmetic formulae in determining need for assistance.

PRESIDENTIAL AND GUBERNATORIAL DISCRETION

The declaration process contains many factors for consideration and, for all but the most catastrophic events, the process moves at a deliberate speed

accumulating information from several sources. While the process is informed by that information and its relationship to potential assistance programs, the information that is gathered at the state and local level does not preclude the exercise of judgment by the governor or the President.

The Stafford Act stipulates several procedural actions a governor must take prior to requesting federal disaster assistance (including the execution within the state of the state emergency plan and an agreement to accept cost-share provisions and related information-sharing). Still, the process leaves broad discretion with the governor if he or she determines that a situation is "beyond the capabilities of the state." The concession that a state can no longer respond on its own is difficult to quantify. It is the governor who makes that assessment, based on his or her knowledge of state resources and capabilities.

Actions by a governor are a driving constant in this process. Both declarations of major disaster and declarations of emergency must be triggered by a request to the President from the governor of the affected state.[27] The President cannot issue either an emergency or a major disaster declaration without a gubernatorial request. The only exception to this rule is the authority given to the President to declare an emergency when "he determines that an emergency exists for which the primary responsibility for response rests with the United States because the emergency involves a subject area for which, under the Constitution or laws of the United States, the United States can exercise exclusive or preeminent responsibility and authority."[28]

The importance of governors in the process is not lost on, nor would it likely be diminished by, Presidents who formerly served in that position. Over the last three decades of Stafford Act implementation, four of the Presidents during this period were former governors who had worked through the disaster declaration process from both the state and the federal level.[29]

Having that experience may have left the Presidents, and their staffs and appointees, with an appreciation of the discretionary authority inherent in the process. While there are some established standards in the law, the factors that are to be weighed in considering the impact of a disaster on the need for assistance for families and individuals are general considerations that underscore the judgment required to reach a decision. As one observer noted of the decision for general, flexible considerations:

> In other words, Congress likes to keep the process imprecise, even if benefits occasionally go to the undeserving. The absence of objective criteria preserves wide political discretion to the president.[30]

Congress also has among its number former governors who have exercised this discretion at the state level. Still, despite the interest some may have in keeping "the process imprecise," some Members of Congress express disappointment at times with the exercise of the discretion and the general nature of the considerations. As noted previously in this report, that disappointment has been reflected in hearings that have focused on how the disaster declaration process works in practice.

In the declaration process, FEMA develops a recommendation that is sent to the White House for action. However, as implied, it is a recommendation from FEMA and the Department of Homeland Security (DHS). The final action, as defined in the regulations of the process, is a "Presidential determination."[31] Just as the governor retains the discretion to request federal assistance regardless of thresholds or indicators, the President retains the discretion to make a decision that may be counter to recommendations he receives.

PRELIMINARY DAMAGE ASSESSMENTS

Although not explicitly mentioned in the Stafford Act, Preliminary Damage Assessments (PDAs) are a crucial part of the process of determining if an event may be declared a major disaster by the President. The minimal discussion of PDAs in the public record stands in inverse proportion to their impact on disaster decisions and subsequent expenditures from the Disaster Relief Fund (DRF).

When a PDA is conducted after an event it is the "mechanism used to determine the impact and magnitude of damage and the resulting unmet needs of individuals, businesses, the public sector, and the community as a whole."[32] The most "preliminary" part of a PDA may be an abbreviated one completed only by the state to determine if the situation merits development of a complete PDA with federal participation. Based on their previous experience, states may determine that the event will not reach the level where a federal disaster declaration is likely. However, despite findings that federal aid may not be needed, there may be political considerations that could lead to a gubernatorial request. As one author points out in his study of the process:

> Governors also feel the heat of media coverage of incidents in their states, and they too appreciate the importance of exhibiting political responsiveness. Governors also appreciate that their future political

> fortunes may be influenced by how they handle their disaster and emergency incidents. As a consequence, the hypothesis assumes that governors are the pivotal and decisive players in securing presidential disaster declarations and that they have a tendency to request declarations for even marginal events.[33]

While media and political pressure may have some influence on the outcome of some requests, governors may exercise caution since they are reluctant to be turned down when requesting aid. A denial of their request could be perceived by some to reflect adversely on their decision-making skills and judgment under pressure. Unlike the procedures of the federal process, a governor's decision to request a declaration is a public and often newsworthy action.

Regardless of any question regarding motivation, the governor's first decision is whether the incident is severe enough to assemble a traditional PDA team to survey the damaged area. The traditional PDA team includes a state official, representatives from the appropriate FEMA regional office, a local official familiar with the area and, in some instances, representatives from the American Red Cross and/or the Small Business Administration. The FEMA representatives have the responsibility of briefing the team on the factors to be considered, the information that will be helpful in the assessment and how the information should be reported. One significant improvement in this process is that the regulations now require that the participants reconcile any differences in their findings.[34]

Another factor is the quality of the PDA team and its findings. PDAs are ordered up quickly after an event. FEMA's 10 regional offices are often engaged in multiple disasters and have to rely on temporary employees (albeit, usually experienced ones) to staff the PDA team.[35] Also, given the variables among regions in interpreting policy and guidance, consistency of approach may also be a question. This became a significant issue during the hurricane season of 2004 in Florida, where questions were raised regarding the designations of some counties.[36]

When working openly and with federal and state cooperation, the PDA can be an effective and inclusive process. As Washington State Emergency Management Director Jim Mullen, and FEMA Region X Director Susan Reinertson, explained the process for a survey of flood damage in November of 2006:

> 'Joint PDA teams will visit and inspect damaged areas, document damage and talk, as needed, with homeowners and local officials,' said

> Mullen. 'It's a partnership effort designed to provide a clear picture of the extent and locations of damage in counties that have reported the most substantial damage to primary homes and businesses.'
>
> 'The PDA teams look at the total scope of damage to establish if recovery is beyond the capabilities and resources of the state and local governments. The PDA doesn't determine the total cost of recovery, nor does it guarantee a presidential declaration for individual assistance,' said FEMA Regional Director Susan Reinertson.[37]

PDA teams often face challenges in the collection of data. Some information may be observable in a survey of the area, such as the number of bridges damaged or the number of culverts washed out. But other necessary information, such as the percentage of elderly residents in an area, or the amount of insurance coverage for all homeowners or renters may be more difficult to obtain. Also the geographic span of the damage can create complications as the PDA team struggles to cover all of the affected area in a limited time. Further complications may then ensue based on how much of the area that the PDA teams have visited (generally counties) are included in the governor's request.

While PDAs are the usual way damages are assessed, there are exceptions to this rule. Some incidents are so massive in their scale and impact that the actual declaration is not in doubt. This would include events such as Hurricanes Andrew and Katrina, the Loma Prieta and Northridge earthquakes, the Mount St. Helens volcanic eruption, and the September 11 terrorist attacks. In these instances, the decision is not whether a declaration will be made, but how broad the coverage will be, both geographic and programmatic. In such cases the President, in accordance with the regulations, can waive the PDA requirement. But as the regulations note, a PDA may still be needed "to determine unmet needs for managerial response purposes."[38] This means the PDA helps to identify a specific, potential need for certain programs, such as crisis counseling or disaster unemployment assistance during the disaster recovery period. It is this identification of discrete need that helps the governor decide on which assistance programs will be requested.

The PDA is a "bottom up" process as information gradually rises up for decision-makers to consider. Multiple pressure points, including affected citizens, elected officials, and professionals in various fields, may all urge PDA team members to reach certain conclusions. As one author suggests:

> The president, motivated by the need to appear highly politically responsive, solicits and encourages a gubernatorial request for a

presidential disaster declaration. Publicity is a factor in that CNN and other news organizations help to promote nationally what would otherwise be a local incident addressed by subnational authorities. The president may also be influenced by the electoral importance of the state that experiences the incident.[39]

It is difficult to overstate the importance of the media context as noted above. Depending on the news of the day, the disaster event in question may be the biggest national story and thus create momentum for action that is difficult to assuage with explanations of traditional administrative procedures.

FACTORS CONSIDERED FOR PUBLIC ASSISTANCE IN MAJOR DISASTER DECLARATIONS

Public Assistance (PA) refers to various categories of assistance to state and local governments and non-profit organizations. Principally, PA covers the repairs or replacement of infrastructure (roads, bridges, public buildings, etc.) but also includes debris removal and emergency protective measures which cover additional costs for local public safety groups incurred by their actions in responding to the disaster.[40] In assessing the degree of PA damage, FEMA considers six general areas:

- Estimated cost of the assistance
- Localized impacts
- Insurance coverage
- Hazard mitigation
- Recent multiple disasters
- Programs of other federal assistance.

Estimated Cost of the Assistance

Although all of these factors are considered, the *estimated cost of the assistance* is a key component and may be "more equal" than other factors since it contains a threshold figure. What is especially noteworthy about the cost estimates is that they could be interpreted as an example of the "arithmetic formula" that was precluded from use in Section 320 of the Stafford Act. However, that section does state that a formula cannot be "solely" determinative of the fate of a governor's request. Use of the other factors listed arguably justifies FEMA's consideration of the threshold figure of $1 million in PA damage—the first number FEMA expects to see in a request that includes PA as an area of needed assistance. In addition FEMA also considers a state-wide threshold of $1.30 per capita in estimated eligible disaster costs before it will approve a request for PA help.[41]

Depending on the state's population, the per capita threshold may be difficult to reach. For example, the 2000 Census estimated California's population at just under 34 million people.[42] Applying the $1.30 per capita figure, it would require eligible PA damage in California to be close to $41.5 million. California is a large state with a budget and tax base commensurate with its size. Expecting a large state to be able to respond on its own is equating such help, and such amounts, to be within "the capabilities of the state."[43]

Compare that level of eligible damage for California ($41.5 million) with Nevada, a small population state according to FEMA regulations.[44] For Nevada, with a population of just under 2 million people according to the 2000 Census, eligible PA damage of about $2.4 million would make the state potentially eligible for supplemental federal assistance.

There are obvious differences in the populations of these two states; however, both have substantial industries and are growing areas. They are bordering states that are both subject to the threat of earthquake damage.[45] Different measurements have been suggested to try to more accurately capture a state's capacity to respond to disaster events. Some of these measurements are discussed in the "Congressional Considerations for the Declaration Process" section of this report.

Localized Impacts

The next factor used by FEMA to consider a declaration that may include PA help focuses on *localized impacts*. FEMA generally looks for a minimum of $3.27 per capita in infrastructure damage in a county before designating it for PA funding.[46] While such a per capita amount would likely ensure that the local entity (almost always a county) would be included in the list of jurisdictions designated for PA assistance should a declaration be issued, high local levels of damage per capita would not supersede a finding that damages state-wide fell below the statewide threshold amount. However, knowledge of a very large localized impact could weigh on the President's discretion and raise the importance of the *localized impact* factor.

Insurance Coverage

Insurance coverage also is considered in assessing damage. Officials preparing PA estimates deduct the amount of insurance that should have been held by units of governments and nonprofit organizations from the total eligible damage amount. However, this is a complicated assessment with several caveats. A considerable number of states and local governments "self-insure" their investments against some types of disasters and may argue that they would have total liability absent a federal disaster declaration.[47] Also, in the event of some disasters (earthquakes being a prime example), the state insurance commissioner must certify that hazard insurance is both available and affordable. If it is not available, deducting an amount of coverage not realistically available would not be a practical consideration.[48]

In the case of a flood event, it is much simpler for FEMA to access information available on whether flood insurance was available, given that FEMA administers the National Flood Insurance Program (NFIP) and there are program staff in each FEMA region who monitor NFIP participation. This knowledge of flood insurance availability and costs makes it a simpler process for FEMA to deduct the amount of flood insurance that should have been in place from the total potential award for a public structure.[49]

Hazard Mitigation

Hazard mitigation presents a challenging and different type of consideration for Public Assistance disaster declaration requests. If the requesting state can prove that their per capita amount of infrastructure damage falls short due to mitigation measures that lessened the disaster's impact, FEMA will consider that favorably in its recommendation to the President. Calculations of savings would be dependant on cost-benefit analysis and other related estimates of damages that were avoided. This factor is intended to encourage mitigation projects by states to lessen the risks of future natural disasters. Some might consider this practice to be one where "a good deed should go unpunished." However, another factor to be considered is that the mitigation work that is credited to the state may, in fact, have been principally financed (up to 75% of the costs) with previous FEMA disaster assistance funding through the Hazard Mitigation Grant Program (HMGP).[50]

Recent Multiple Disasters

Recent multiple disasters is the factor FEMA considers when a state has been repeatedly hit by disaster events (either presidential declarations or events within the state that were not declared) within the previous 12 months. FEMA evaluates the amount of funds that the state has committed to these recent events and their impact on the state and its residents. For example, a request from a state that has responded on it own to a series of tornadoes may receive a more favorable consideration, even if the catalyst for the request was arguably not as destructive as others. This factor was used recently in FEMA's assessment of the multiple hurricanes that struck Florida during the 2004 hurricane season.

Other Federal Programs

When FEMA is reviewing a governor's request it is also considering whether *other federal programs* are available. One example might be if a large amount of the reported damage occurred to federal-aid-system roads. The Federal Highway Administration (FHWA) is a more appropriate agency to handle such an event, and administers programs that address this specific type of damage.[51] The bridge collapse in Minnesota in August of 2007 was a dramatic example of this type of event that would warrant significant, non-FEMA, federal aid. Similarly, other federal programs may be more responsive to certain types of natural events and the problems they create. For example, oceanic bacteria could cause harmful failures for the fishing and hatcheries industry. Should such outbreaks occur, the Magnuson-Stevens Fisheries Act (P.L. 94-265) authorizes programs under the Commerce Department that would have more appropriate forms of emergency assistance for commercial fishermen than FEMA through the Stafford Act.[52] Also, while drought is a type of disaster that could result in a major disaster, such catastrophes usually result in program assistance from the U.S. Department of Agriculture rather than presidential declarations.[53]

In summary, all of these factors are used by FEMA to determine whether a major disaster declaration will be recommended and whether PA aid will be extended as a part of that potential declaration. FEMA's assistance to a homeowner who chooses not to purchase insurance is left with help from FEMA that is capped at $28,200. The assistance to repair public infrastructure is based on the amount of eligible damage caused by the disaster event. There is no cap, and the sums can be in the billions of dollars.

FACTORS CONSIDERED FOR INDIVIDUAL ASSISTANCE IN MAJOR DISASTER DECLARATIONS

Individual Assistance (IA) includes various forms of help for families and individuals following a disaster event. The assistance authorized by the Stafford Act can include housing assistance, disaster unemployment assistance, crisis counseling and other programs intended to address the needs of people. In seeking to assess the impact of a disaster on families and individuals, the factors FEMA considers include:

- Concentration of damages
- Trauma
- Special populations
- Voluntary agency assistance
- Insurance coverage
- Average amount of Individual Assistance by state.

Concentration of Damages

Concentration of damages looks at the density of the damage in individual communities. FEMA's regulations state that highly concentrated damages "generally indicate a greater need for federal assistance than widespread and scattered damages throughout a state."[54] Concentrated damages are far more visible and may be an indication of significant damage to infrastructure supporting neighborhoods and communities, thereby increasing the needs of individuals and families.

However, the dispersion of damage is not necessarily an indication that individual and family needs are non-existent. Damage in rural states, almost by definition, is far less concentrated and could arguably be more difficult for a PDA team to view and assess. Congress has sought to address this challenge through the creation of the Rural and Small State advocate position at FEMA.[55]

Trauma

Trauma is defined in three ways in FEMA's regulations: the loss of life and injuries, the disruption of normal community functions, and emergency needs that could include an extended loss of power or water.[56]

Despite their prominence and importance to victims, families and communities, the loss of life and injuries have relatively little bearing on a declaration decision, but they would greatly influence media coverage that can influence the decision-making process.

An extreme amount of losses would be traumatic and considered in the evaluation of the governor's request. But the actual help available from FEMA in response to such losses is limited. The Other Needs Assistance (ONA) program, a part of the Individuals and Households Program (IHP), may provide assistance for uninsured funeral expenses and medical help.[57]

As noted, the extreme loss of life or many injuries will likely influence the amount of media coverage for the event. As previously explained in the "Presidential and Gubernatorial Discretion" section of this report, the media coverage can influence not merely the pace of the decision, but the actual decision itself.[58]

The other areas (disruption of functions and loss of utilities) in the trauma rubric can be a point of dispute in declaration decisions. What constitutes a disruption of normal community functions?

Do road closures that result in the closing of schools equate to a disruption? Another consideration is how long the disruption remains and how, or even if, federal help can alleviate the disruption. Similarly, the emergency needs due to the loss of utilities are defined by the length of time the power or water are not in service. Because characteristics of these factors are not defined in regulations, discretionary judgments are significant aspects of the evaluation of IA needs.

Special Populations

Special populations are considered by FEMA in assessing a request for Individual Assistance.

FEMA attempts to ascertain information about the demographics of an area affected by a disaster event. Those demographics include the age and income of residents, the amount of home ownership in an area, the effect on Native American tribal groups, and other related considerations to be taken into account. The knowledge of the demographics within the affected area gives FEMA added information to consider regarding trauma and community disruption. Special populations are a factor in the consideration of a governor's request, but the total number of households affected that would be eligible for Stafford Act programs remains the prime consideration.

Voluntary Agency Assistance

Voluntary agency assistance involves an evaluation of what the volunteer and charitable groups and state and local governments have already done to assist disaster victims as well as the potential help they can offer in the recovery period. FEMA also considers whether state or local programs "can meet the needs of disaster victims."[59] This factor is among the most contentious in the disaster declaration process due to the subjectivity of the assessment.

FEMA's evaluation of local and state capabilities, and the capabilities of the local voluntary community, may vary greatly among catastrophes. There is an expectation that the regional office's relationship and history with the state could provide some of this information. But a cursory look at state programs available to "meet the needs of disaster victims" suggests that few resources are comparable to federal help in intent or in scope. A National Emergency Management Association (NEMA) survey of its members for state-funded disaster assistance showed that relatively few states provide assistance beyond that authorized in the Stafford Act.

NEMA noted that the "other" category could include a range of programs or funds not necessarily of direct aid to victims, including "local government loans, HAZMAT (hazardous materials) funds, disaster unemployment insurance, and a governor's contingency fund."[60] The report does not define these programs any further, but the small numbers of programs directed at IA or unmet needs suggests that FEMA officials evaluate this area carefully given the lack of information regarding available resources for these programs and the extent of their coverage.[61]

Table 1 summarizes data from that survey

Assistance Provided	Number of States with These Programs
Public Assistance Program	22
Individual Assistance Program	9
Unmet Need Program	1
Other	5

Source: NEMA 2010 Biennial Report: Organizations & Funding for State Emergency Management and Homeland Security (Lexington, KY: Council of State Governments) p. 4.

In the same vein, attempts to assess the capacity of local voluntary and charitable groups to handle "unmet needs" caused by a disaster can be challenging and problematic. Part of the challenge is discerning the assistance available from the non-profit, voluntary sector, and if that aid meets the needs created by the disaster event. A problematic aspect of the assessment is similar to the "good deeds" concept addressed in the "hazard mitigation" factor in PA. In that instance, the criteria reward the mitigation work that communities and states had undertaken to lessen the impact of a disaster. Similarly, a strong and efficient charitable sector at the local level that is equipped and funded to address the remaining needs could result in a disaster not being declared and federal supplemental funding not being made available.

Arguably, a community does not base all of its preparedness decisions on the potential of FEMA funding in an extraordinary situation. However, it has been argued that some aspects of the FEMA declaration process could be viewed as disincentives for a sound, local capacity to deliver such assistance. One analyst believed that this problem underlines the need for clear criteria for smaller disaster events:

> Too low a threshold reinforces the perception that the federal government will always come like the cavalry to rescue states and local governments from their improvident failure to prepare for routine disasters. Lapses in preparedness, response, recovery, and mitigation (to cite the disaster management litany) should not be encouraged by a too readily available bailout by the federal government and taxpayers.62

In evaluating the capabilities of local organizations, FEMA seeks to determine the help that actually exists, rather than assistance that might be expected to be in place.

Insurance Coverage

As noted earlier in the PA section, *insurance coverage* is also an important consideration when FEMA considers a request for Individual Assistance. This is in part derived from the general prohibition in the statute of the duplication of benefits, as follows:

> ... each federal agency administering any program providing financial assistance to persons, business concerns, or other entities suffering losses as a result of a major disaster or emergency, shall assure that no such person, business concern or other entity will receive such assistance with respect to any part of such loss as to which he has received financial assistance under any other program or from insurance or any other source.[63]

This provision does not necessarily result in delayed assistance. FEMA is able to provide help to individuals and households that have disaster damages but are waiting on insurance or other assistance for help. Those applicants can receive FEMA help as long as they agree to reimburse FEMA when they receive their other assistance. If a disaster occurred where almost all of the damaged dwellings were fully insured for the damage that was sustained, FEMA could conclude that a disaster declaration by the President was not necessary. Among the types of disasters FEMA frequently responds to, tornado disasters particularly reflect this challenge since tornado coverage is a part of most homeowners insurance policies.

Since the NFIP is administered by FEMA, officials can quickly determine the status of flood insurance in communities and the number of policies in place in the affected area. Additionally, knowledge of the income of the area's residents, as suggested in the *special populations* factor, allows the agency to make some projections regarding the likelihood of insurance coverage, particularly special hazard insurance such as flood or earthquake insurance, which are potentially expensive additions to a homeowners policy. As one analyst has noted:

> The decision not to buy insurance for earthquakes, floods, hurricanes and other natural disasters isn't always conscious: some homeowners don't realize they're not already covered. But many others, faced with high premiums and policies with limited coverage, gamble that they won't need insurance help to rebuild after a disaster.64

Average Amount of Individual Assistance by State

The last factor FEMA considers in assessing IA needs is the *average amount of Individual Assistance by state.* FEMA has issued statistics on average losses but notes that the average numbers used are not a threshold (see *Table 2*). The agency does suggest that the "following averages may prove useful to states and voluntary agencies as they develop plans and programs to meet the needs of disaster victims." The inference is that the levels listed generally are what would be expected in damage to dwellings.

Table 2. Average Amount of Federal Assistance Per Disaster Based on Size of State, July 1995 - July 1999

Categories	Small States <2MillionPopulation	Medium States 2-10 Million Population	Large States >10 Million Population
Avg. population (1990 Census Data)	1,000,057	4,713,548	15,552,791
Number of disaster housing applications approved	1,507	2,747	4,679
Number of homes est. major damage/destroyed	173	582	801
Dollar amount of housing assistance	$2.8 million	$4.6 million	$9.5 million
Number of Individual and Family Grants (IFG) approved (Now known as ONA)	495	1,377	2,071
Dollar amount of IFG assistance (now ONA)	$1.1 million	$2.9 million	$4.6 million
Disaster housing/IFG (ONA) combined assistance	$3.9 million	$7.5 million	$ 14.1 million

Source: 44 CFR §206.48(b) (6).

The average numbers that follow are based on disasters that occurred between July of 1995 and July of 1999; the data are 10 years old and of questionable use today. The chart divides states into three categories: small states (under 2 million in population), medium states (2 to 10 million in population) and large states (over 10 million in population). The population amounts are based on the 1990 Census.

While FEMA's regulations stress that these are not thresholds, they are considered by agency officials in determining whether IA will be provided. Presumably states may consider that FEMA help could be forthcoming if damage reaches the state indicator levels. However, since the amounts have not been updated in ten years and are based on a 1990 census, it is difficult to determine the degree to which these numbers are considered. Given Congress's mandate in Section 320 of the Stafford Act, this cannot be an arithmetic formula that solely determines whether assistance is provided. But the presentation of loss indicators may guide states when considering whether to request assistance.

In addition to the issue of the data's currency, a larger and more compelling question is whether the numbers in the chart match up to the Preliminary Damage Assessments (PDAs) for those events. Absent an existing review of detailed information in PDA forms, it is not possible to determine the usefulness of the data in *Table 2*.[65] Still, given the level of experience in this field following literally thousands of disaster declarations over the last 30 years, it could be argued that the numbers of eligible households assisted may be reflective of the estimated damage upon which the decision for a disaster declaration was made.

Congressional Considerations for the Declaration Process

When Congress considers the mandated GAO reports and other commentary on the declaration process, there are some considerations that Members may wish to review when considering the current declaration process.

The Composition of Preliminary Damage Assessment Teams

One area of consideration is the composition of Preliminary Damage Assessment teams. Team members are involved throughout the process and include local officials guiding the team, state personnel who assist the governor in requesting assistance, and FEMA staff who work on the disaster if one is declared. While FEMA staff have the opportunity at several levels to refine the information the team gathers on damages and to ask additional questions, the process could be approached in other ways. Some FEMA staff have suggested the formation of several permanent teams that would have PDAs as their prime job task without continuing involvement in particular disasters. This could result in more consistent assessments with the bonus of added perspective of team members with exposure to various disasters in many regions of the country. Establishing independent and expert on-site groups to review a situation is a recommendation that the National Transportation Safety Board (NTSB) has been contemplating in their investigations of transportation accidents.[66]

The PDA is an important part of the declaration process. While it can be subject to challenge, it is also the assessment closest to the event. Given its vital role in the process the PDA deserves close attention to determine if these on-site assessments are accurately reflecting the character of an event and the likely eligible damages following a disaster event.

Updating and Revising Individual Assistance Averages

As previously noted, the averages that appear in FEMA's declaration process regulations for Individual Assistance (IA) to states are derived from experiences from July 1995 to July 1999.[67] These figures could be updated based on FEMA's more recent experience in delivering this type of assistance. Legislation introduced in the 111th Congress had a provision that directed FEMA to review the factors that are considered for an IA designation.[68]

While the averages are separated by state population size, there does not appear to be a threshold number equivalent to the PA figure of $1 million in eligible damage. In the case of IA help, a dollar figure may not be desirable; however, numbers of families and households affected, or a minimum number of homes with major damage, could potentially be a starting point in establishing a base IA threshold for consideration of requests.

Other Potential Disaster Indicators

The current per capita indicator based on state population according to the U.S. Census is clear, but some observers believe it lacks precision. For example, the earlier discussion on PA indicators pointed out the commonality between California and Nevada regarding earthquake risk as well as the growth of the states. Their population sizes are very different but the per capita indicator alone does not necessarily measure a state's fiscal capacity.

The GAO report in 2001 noted that per capita personal income measures do not take into account the taxable income a state may enjoy from businesses or corporations that may have considerable taxable profits. GAO did suggest a different indicator:

> We have previously reported that Total Taxable Resources (TTR), a measure developed by the U.S. Department of Treasury, is a better measure of state funding capacity in that it provides a more comprehensive measure of the resources that are potentially subject to state taxation.[69]

FEMA's response to this suggestion questioned whether the use of TTR would be in violation of the legislative prohibition against arithmetic formulas.[70] The TTR does have very specific information and also tracks growth within a state. However, this appears to make it a more accurate measurement. The degree of detail in the measurement would not necessarily make it the "sole" determinant of disaster aid. It could remain an indicator and one among several factors to be considered in the declaration process.

Concluding Observations

The disaster declaration process, though subject to inquiry, argument, hearings, studies and recommendations, has changed very little over time. It remains a process that can be observed and evaluated as it occurs in the area affected by the disaster, and grows opaque as it moves up through layers of FEMA and DHS management to the White House. By making PDA information available, FEMA has begun to lift the veil on the decisions that are made. Congress has demonstrated an interest in this process and has sought to understand its limits and its effectiveness. "From the Government Accountability Office in 1981 to Vice President Gore's National Performance Review in 1994," one student of the process noted, "countless policy critiques have called for more objective criteria for presidential disaster declarations."[71] But those calls can be muted when a disaster occurs in a particular place that has particular importance to actors in the process, whether in Congress or the executive branch.

While criteria have been established to create a relatively uniform and knowable process, these criteria may not be determinative of the most critical elements considered. An emphasis on victims of unexpected natural disaster events will likely always have a compelling influence on the disaster declaration process. But since disaster relief does have a political element, at many levels, precedent arguably can have value in improving the declaration process so that it can be applied broadly and fairly. But precedent can also be problematic when discussing events of very different size and impact.

During the last Congress, attention was directed to the nature of the disaster declaration process. The 112th Congress may choose to consider a broad review of the process that might include the consistency of FEMA's approach across the nation in making damage assessments, other potential indicators of state capabilities and capacities, and the currency of the factors it employs to evaluate those assessments of disaster damage and the state requests on which they are based.

End Notes

[1] 42 U.S.C. § 5122.

[2] Ibid.

[3] For additional discussion, see CRS Report RL34724, *Would an Influenza Pandemic Qualify as a Major Disaster Under the Stafford Act?*, by Edward C. Liu.

[4] For more information on disaster supplementals, see CRS Report RL33226, *Emergency Supplemental Appropriations Legislation for Disaster Assistance: Summary Data*, by Justin Murray and Bruce R. Lindsay. For a map of disaster declarations, see http://www.bakerprojects.com.fema/Maps/CONUS.pdf.

[5] Rutherford H. Platt, *Disasters and Democracy: The Politics of Extreme Natural Events*, (Washington, DC: Island Press, 1999), p. 66.

[6] 42 U.S.C. § 5170.

[7] 42 U.S.C. § 5191.

[8] Michael J. Trebilcock and Ronald J. Daniels, "Rationales and Instruments for Government Intervention in Natural Disasters" in Ronald J. Daniels, Donald F. Kettl, and Howard Kunreuther, eds., *On Risk and Disaster: Lessons from Hurricane Katrina*, (Philadelphia: University of Pennsylvania Press, 2006), p. 104. For additional discussion of the evolution of emergency management see Claire B. Rubin, ed. *Emergency Management: The American Experience 1900-2005* (Fairfax, VA: Public Entity Risk Institute, 2007).

[9] U.S. Department of Homeland Security, Federal Emergency Management Agency, Annual Report of FOIA Activity for FY2000, II. C., at http://www.fema.gov/doc/help/foiareportfy00.doc.

[10] U.S. Department of Homeland Security, Federal Emergency Management Agency, *Preliminary Damage Assessment Reports,* at http://www.fema.gov/rebuild/recover/pda-reports.shtm.

[11] Ibid.; and Platt, p. 266.

[12] U.S. Government Accountability Office, *Disaster Assistance: Improvements Needed in Disaster Declaration Criteria and Eligibility Assurance Procedures,* GAO-01-837, August, 2001, p. 2.

[13] For more information on emergency supplemental spending for disasters see CRS Report RL33226, *Emergency Supplemental Appropriations Legislation for Disaster Assistance: Summary Data*, by Justin Murray and Bruce R. Lindsay.

[14] U.S. Congress, House Committee on the Budget, *Budgetary Treatment of Emergencies,* p. 68, 105th Congress, 2nd. Sess., June 23, 1998 (Washington: GPO, 1998).

[15] For more information, see CRS Report RL33053, *Federal Stafford Act Disaster Assistance: Presidential Declarations, Eligible Activities, and Funding*, by Keith Bea.

[16] Ibid.; and Platt, p. 10.

[17] U.S. Congress, Senate Committee on Appropriations, Subcommittee on VA, HUD and Independent Agencies, *Departments of Veterans Affairs and Housing and Urban Development and Independent Agencies Appropriations for Fiscal Year 2002,* 107th Congress, 1st sess., p. 252, at http://www.access.gpo.gov/congress/senate.

[18] Opening Statement of Rep. Bennie Thompson, in U.S. Congress, House Committee on Homeland Security, *Disaster Declarations, Where is FEMA in a Time of Need?,* 110th Congress, 1st. sess., March 15, 2007.

[19] P.L. 110-28, 121 Stat. 155.

[20] U.S. Congress, House Committee on Appropriations, *Making Emergency Supplemental Appropriations for the Fiscal Year ending September 30, 2007 and other purposes, report to accompany H.R. 1591*, 110th Cong., 1st sess., March 20, 2007, H.Rept. 110-60, p. 211.

[21] Government Accountability Office, *FEMA Disaster Cost Estimates,* Feb. 2008, p. 4.

[22] Matt Korade, "FEMA's Success Hinges on Regional Personnel," *Congressional Quarterly,* March 16, 2010, at http://homeland.cq.com/hs/display/do?docid=3569657&sourcetype=31.

[23] P.L. 109-295, 689g, 120 Stat. 1453.

[24] For more information see CRS Report RL33729, *Federal Emergency Management Policy Changes After Hurricane Katrina: A Summary of Statutory Provisions*, coordinated by Keith Bea.

[25] U.S. Congress, House Subcommittee on Investigations and Oversight, *The Federal Emergency Management Agency's Proposed Disaster Relief Regulations,* 100th Cong., 1st sess. (Washington: GPO, 1987), pp. 3-5.

[26] 42 U.S.C. 5163.

[27] For more information, see CRS Report RL33090, *Robert T. Stafford Disaster Relief and Emergency Assistance Act: Legal Requirements for Federal and State Roles in Declarations of an Emergency or a Major Disaster*, by Elizabeth B. Bazan.

[28] 42 U.S.C. §5191.

[29] President Carter (Georgia), President Reagan (California), President Clinton (Arkansas), and President George W. Bush (Texas).

[30] Ibid.; and Platt, p. 20.

[31] CFR 44 § 206.38.

[32] CFR 44 § 206.33.

[33] Richard Sylves, Quick Response Report #86: *The Politics and Administration of Presidential Disaster Declarations: The California Floods of Winter 1995,* at http://www.colorado.edu/hazards/research/qr/qr86.html.

[34] 44 CFR §206.33(c).

[35] FEMA has 10 regional offices: Region 1 (Boston, MA), Region 2 (New York, NY), Region 3 (Philadelphia, PA), Region 4 (Atlanta, GA), Region 5 (Chicago, IL), Region 6 (Denton,

TX), Region 7 (Kansas City, MO), Region 8 (Denver, CO), Region 9 (Oakland, CA), and Region 10 (Bothell, WA).

[36] Sally Kestin and Megan O'Matz, "FEMA ruled on disaster before verifying Dade damage," *South Florida Sun-Sentinel*, at http://www.sun-sentinel.com/news/sfl-fema15may15,0,6848867.story?coll=sfla-news-utilities.

[37] U.S. Department of Homeland Security, Federal Emergency Management Agency, "Joint State-Federal Preliminary Damage Assessment Teams Visit Flood-Damaged Communities," Release Number R10-06-047, November 13, 2006, at http://www.fema.gov/news/newsrelease.fema?id=31509.

[38] 44 CFR 206.33(d)

[39] Ibid.; and Sylves, p. 5.

[40] U.S. Department of Homeland Security, Federal Emergency Management Agency, *Presidential Declarations: What Does This Mean for Me?,* at http://www.fema.gov/media/archives/2007/010807b.shtm.

[41] U.S. Department of Homeland Security, Federal Emergency Management Agency, "Notice of Adjustment of Statewide Per Capita Impact Indicator," 75 *Federal Register,* 62135, October 7, 2010.

[42] U.S. Census Bureau, *State & County Quick Facts,* at http://quickfacts.census.gov/gfd/states/06000.html.

[43] 42 U.S.C. §5170.

[44] 44 CFR § 206.48.

[45] U.S. Geological Survey, Earthquake Hazards Program, "Earthquake Density Maps for the United States", at http://earthquake.usgs.gov/regional/states/us_density.php.

[46] U.S. Department of Homeland Security, Federal Emergency Management Agency, "Notice of Adjustment of Countywide Per Capita Impact Indicator," 75 *Federal Register,* 62135, October 7, 2010.

[47] For information on public insurance coverage, see "Cost of Risk Survey" conducted by the Public Entity Risk Institute (PERI) at http://www.riskinstitute.org.

[48] 44 CFR §206.253. For additional information, see CRS Report RS22945, *Flood Insurance Requirements for Stafford Act Assistance*, by Edward C. Liu.

[49] 44 CFR §206.252.

[50] 42 U.S.C. § 5170c.

[51] For more information, see CRS Report RS22268, *Repairing and Reconstructing Disaster-Damaged Roads and Bridges: The Role of Federal-Aid Highway Assistance*, by Robert S. Kirk.

[52] For additional discussion of this law, see CRS Report RL34209, *Commercial Fishery Disaster Assistance*, by Harold F. Upton.

[53] For more information, see CRS Report RS21212, *Agricultural Disaster Assistance*, by Dennis A. Shields and Ralph M. Chite.

[54] 44 CFR §206.48(b)(1).

[55] P.L. 109-295, §689g, 120 Stat. 1453.

[56] 44 CFR §206.48(b)(2).

[57] 42 U.S.C. § 5174.

[58] Richard Sylves, Quick Response Report #86: *The Politics and Administration of Presidential Disaster Declarations: The California Floods of Winter 1995*, at http://www.colorado.edu/hazards/research/qr/qr86.html

[59] 44 CFR §206.48 (b) (4).

[60] National Emergency Management Association (NEMA) 2006 Biennial Report: Organizations and Funding for State Emergency Management and Homeland Security, (Lexington, KY: Council of State Governments).

[61] For more information on state assistance programs, refer to the following for summary information on state statutory authorities: CRS Report RL32287, *Emergency Management and Homeland Security Statutory Authorities in the States, District of Columbia, and Insular Areas: A Summary*, by Keith Bea, L. Cheryl Runyon, and Kae M. Warnock.

[62] Ibid.; and Platt, p. 65.

[63] 42 U.S.C. §5155.

[64] Liz Pulliam Weston, "Do You Really Need Disaster Insurance?," *MSN Money,* at http://moneycentral.msn.com/content/Insurance/Insureyourhome/P59648.hsp?Printer.

[65] This statement remains accurate. However, since FEMA has begun to list PDA information on its website, it should be possible to begin to assess the efficacy of the factors employed to determine IA damage and declaration criteria.

[66] For more information, see CRS Report RL33474, *Reauthorization of the National Transportation Safety Board (NTSB)*, by Bart Elias.

[67] 44 CFR § 206.48.

[68] H.R. 3377, Title III, Sec.304.

[69] U.S. Government Accountability Office, *DISASTER ASSISTANCE: Improvement Needed in Disaster Declaration Criteria and Eligibility Assurance Procedures,* GAO-01-837, August 2001, p.11.

[70] Ibid., p. 49.

[71] Ibid.; and Platt, p. 285.

INDEX

D

E

O

P

R

W